ENERGY POLICIES, POLITICS AND PRICES SERIES

U.S. ENERGY: OVERVIEW OF THE TRENDS, STATISTICS, SUPPLY AND CONSUMPTION

ENERGY POLICIES, POLITICS AND PRICES SERIES

Nuclear Power's Role in Generating Electricity
Perry G. Furham
2009. ISBN: 978-1-60741-226-7

OPEC, Oil Prices and LNG
Edward R. Pitt and Christopher N. Leung (Editors)
2009. ISBN: 978-1-60692-897-4

OPEC, Oil Prices and LNG
Edward R. Pitt and Christopher N. Leung (Editors)
2009. ISBN: 978-1-60876-614-7
(Online Book)

Energy Prices: Supply, Demand or Speculation?
John T. Perry (Editor)
2009. ISBN: 978-1-60741-374-5

Dynamic Noncooperative Game Models for Deregulated Electricity Markets
Jose B. Cruz, Jr. and Xiaohuan Tan
2009. ISBN: 978-1-60741-078-2

Power Plant Characteristics and Costs
Stan Kaplan
2009. ISBN: 978-1-60741-264-9

Hydrogen Fuel Perspectives
Ian S. Rubio (Editor)
2009. ISBN: 978-1-60692-444-0

Natural Gas Markets and Lessons Learned
E. K. Cho (Editor)
2010. ISBN: 978-1-61668-249-1

Natural Gas Markets and Lessons Learned
E. K. Cho (Editor)
2010. ISBN: 978-1-61668-429-7
(Online Book)

Strategic Petroleum Reserve
Albert L. Strait (Editor)
2010. ISBN: 978-1-60692-290-3

Solar Energy Technologies: From Research to Deployment
Liam G. White (Editor)
2010. ISBN: 978-1-60741-323-3

U.S. Energy: Overview of the Trends, Statistics, Supply and Consumption
Gregor E. Peake (Editor)
2010. ISBN: 978-1-60876-041-1

**Federal Energy Management
and Government Efficiency Goals**
*Amelia R. Williams
(Editor)*
2010. ISBN: 978-1-60692-985-8

**Federal Energy Management
and Government Efficiency Goals**
Amelia R. Williams (Editor)
2010. ISBN: 978-1-61668-538-6
(Online Book)

**Rational Use and Energy Planning:
A Thermodynamic
and Geographical Approach**
*Giuseppe Grazzini, Carla Balocco
and Giovan Battista Andreani*
2010. ISBN: 978-1-60741-350-9

World Biofuels Production Potential
Thomas E. Rommer (Editor)
2010. ISBN: 978-1-61668-663-5

World Biofuels Production Potential
Thomas E. Rommer (Editor)
2010. ISBN: 978-1-61668-425-9
(Online Book)

**Transition to Hydrogen
Fuel Cell Vehicles**
Selim Koca (Editor)
2010. ISBN: 978-1-60741-806-1

**Employment Effects of Transition
to a Hydrogen Economy in the U.S.**
*Michele Auriemma
(Editor)*
2010. ISBN: 978-1-60741-808-5

**The Role of Auctions in Emission
Allowance Allocations
for Greenhouse Gases**
*Aubrey D. O'Connor
(Editor)*
2010. ISBN: 978-1-60741-699-9

**The Role of Auctions in Emission
Allowance Allocations
for Greenhouse Gases**
*Aubrey D. O'Connor
(Editor)*
2010. ISBN: 978-1-61668-652-9
(Online Book)

Reducing Greenhouse Gas Emissions
Joseph G. Levitt (Editor)
2010. ISBN: 978-1-60741-890-0

Reducing Greenhouse Gas Emissions
Joseph G. Levitt (Editor)
2010. ISBN: 978-1-61668-730-4
(Online Book)

**Energy Efficiency through
Combined Heat and Power
or Cogeneration**
*David H. Thomas
(Editor)*
2010. ISBN: 978-1-61668-341-2

**Energy Efficiency through
Combined Heat and Power
or Cogeneration**
*David H. Thomas
(Editor)*
2010. ISBN: 978-1-61668-432-7
(Online Book)

**U.S. Energy and the Environment:
An Overview
and Comparative Analysis**
*Roland H. Terrison
(Editor)*
2010. ISBN: 978-1-61668-017-6

**U.S. Energy and the Environment:
An Overview
and Comparative Analysis**
*Roland H. Terrison
(Editor)*
2010. ISBN: 978-1-61668-641-3
(Online Book)

The Smart Grid and Electric Power Transmission
*Caitlin G. Elsworth
(Editor)*
2010. ISBN: 978-1-61668-223-1

Combined Heat and Power - Analysis of Various Markets
Jordan A. Cory
2010. ISBN: 978-1-60741-269-4

Combined Heat and Power - Analysis of Various Markets
Jordan A. Cory
2010. ISBN: 978-1-61668-377-1
(Online Book)

Worldwide Biomass Potential: Technology Characterizations
R. L. Bain
2010. ISBN: 978-1-60741-267-0

The Completion of the Oil Era: The Economic Impact
Carlos A. Rossi
2010. ISBN: 978-1-60741-340-0

ENERGY POLICIES, POLITICS AND PRICES SERIES

U.S. ENERGY: OVERVIEW OF THE TRENDS, STATISTICS, SUPPLY AND CONSUMPTION

GREGOR E. PEAKE
EDITOR

Nova Science Publishers, Inc.
New York

Copyright © 2010 by Nova Science Publishers, Inc.

All rights reserved. No part of this book may be reproduced, stored in a retrieval system or transmitted in any form or by any means: electronic, electrostatic, magnetic, tape, mechanical photocopying, recording or otherwise without the written permission of the Publisher.

For permission to use material from this book please contact us:
Telephone 631-231-7269; Fax 631-231-8175
Web Site: http://www.novapublishers.com

NOTICE TO THE READER

The Publisher has taken reasonable care in the preparation of this book, but makes no expressed or implied warranty of any kind and assumes no responsibility for any errors or omissions. No liability is assumed for incidental or consequential damages in connection with or arising out of information contained in this book. The Publisher shall not be liable for any special, consequential, or exemplary damages resulting, in whole or in part, from the readers' use of, or reliance upon, this material. Any parts of this book based on government reports are so indicated and copyright is claimed for those parts to the extent applicable to compilations of such works.

Independent verification should be sought for any data, advice or recommendations contained in this book. In addition, no responsibility is assumed by the publisher for any injury and/or damage to persons or property arising from any methods, products, instructions, ideas or otherwise contained in this publication.

This publication is designed to provide accurate and authoritative information with regard to the subject matter covered herein. It is sold with the clear understanding that the Publisher is not engaged in rendering legal or any other professional services. If legal or any other expert assistance is required, the services of a competent person should be sought. FROM A DECLARATION OF PARTICIPANTS JOINTLY ADOPTED BY A COMMITTEE OF THE AMERICAN BAR ASSOCIATION AND A COMMITTEE OF PUBLISHERS.

LIBRARY OF CONGRESS CATALOGING-IN-PUBLICATION DATA

U.S. energy : overview of the trends, statistics, supply and consumption / editor, Gregor E. Peake.
    p. cm.
  Includes index.
  ISBN 978-1-60876-041-1 (softcover)
  1. Energy industries--United States--Statistics. 2. Energy consumption--United States--Statistics. 3. Energy development--United States--Statistics. I. Peake, Gregor E. II. Title: US energy.
    HD9502.U52U1724 2009
    333.790973--dc22
                                                                                        2009038422

Published by Nova Science Publishers, Inc., ✝ *New York*

CONTENTS

Preface		ix
Chapter 1	Energy Outlooks and Federal Resources Hearing-Birol Testimony *Fatih Birol*	1
Chapter 2	Energy Outlooks and Federal Resources Hearing-Gruenspecht Testimony *Howard Gruenspecht*	5
Chapter 3	Energy Outlooks and Federal Resources Hearing-Pierce Testimony *Brenda S. Pierce*	15
Chapter 4	U.S. Energy: Overview and Selected Facts and Numbers *Carl E. Behrens and Carol Glover*	25
Chapter Sources		57
Index		59

PREFACE

Chapter 1 - This chapter is written testimony by Dr. Fatih Birol before the Subcommittee on Energy and Mineral Resources, Committee on Natural Resources, U.S. House of Representatives, dated March 5, 2009.

Chapter 2 - This chapter is testimony of Dr. Howard Gruenspecht before the Subcommittee on Energy and Mineral Resources, Committee on Natural Resources, U.S. House of Representatives, dated March 5, 2009.

Chapter 3 - This chapter is testimony Brenda S. Pierce before the Subcommittee on Energy and Mineral Resources, Committee on Natural Resources, U.S. House of Representatives, dated March 5, 2009.

Chapter 4 - Energy supplies and prices are major economic factors in the United States, and energy markets are volatile and unpredictable. Thus, energy policy has been a recurring issue for Congress since the first major crisis in the 1970s. As an aid in policy making, this report presents a current and historical view of the supply and consumption of various forms of energy.

The historical trends show petroleum as the major source of energy, rising from about 38% in 1950 to 45% in 1975, then declining to about 40% in response to the energy crisis of the 1970s. Significantly, the transportation sector has been and continues to be almost completely dependent on petroleum, mostly gasoline. The importance of this dependence on the volatile world oil market was revealed over the past five years as perceptions of impending inability of the industry to meet increasing world demand led to relentless increases in the prices of oil and gasoline. With the downturn in the world economy and a consequent decline in consumption, prices collapsed, but the dependence on imported oil continues as a potential problem.

Natural gas followed a similar pattern at a lower level, increasing its share of total energy from about 17% in 1950 to more than 30% in 1970, then declining to about 20%. Consumption of coal in 1950 was 35% of the total, almost equal to oil, but it declined to about 20% a decade later and has remained at about that proportion since then. Coal currently is used almost exclusively for electric power generation.

Nuclear power started coming online in significant amounts in the late 1960s. By 1975, in the midst of the oil crisis, it was supplying 9% of total electricity generation. However, increases in capital costs, construction delays, and public opposition to nuclear power following the Three Mile Island accident in 1979 curtailed expansion of the technology, and many construction projects were cancelled. Continuation of some construction increased the nuclear share of generation to 20% in 1990, where it remains currently. The first new reactor license applications in nearly 30 years were recently submitted, but no new plants are currently under construction or on order.

Construction of major hydroelectric projects has also essentially ceased, and hydropower's share of electricity generation has gradually declined, from 30% in 1950 to 15% in 1975 and less than 10% in 2000. However, hydropower remains highly important on a regional basis.

Renewable energy sources (except hydropower) continue to offer more potential than actual energy production, although fuel ethanol has become a significant factor in transportation fuel, and wind power has recently grown rapidly. Conservation and energy efficiency have shown significant gains over the past three decades and offer encouraging potential to relieve some of the dependence on imports that has caused economic difficulties in the past, as well as the present.

After an introductory overview of aggregate energy consumption, this report presents detailed analysis of trends and statistics regarding specific energy sources: oil, electricity, natural gas, coal and renewable energy. A section on trends in energy efficiency is also presented.

In: U.S. Energy: Overview of the Trends...
Editors: Gregor E. Peake

ISBN: 978-1-60876-041-1
© 2010 Nova Science Publishers, Inc.

Chapter 1

ENERGY OUTLOOKS AND FEDERAL RESOURCES HEARING-BIROL TESTIMONY

Fatih Birol

Chairman Costa, members of the committee, thank you for the opportunity to appear before you today to discuss the views of the International Energy Agency (IEA) on the outlook for global energy markets over the medium and longer-term. My name is Fatih Birol and I am the Chief Economist and the Director of the office responsible for the economic analysis of energy policy at the IEA.

By way of background, the IEA is an intergovernmental organisation which acts as an advisor to 28 member countries including the United States in their effort to ensure reliable, affordable and clean energy for their citizens. Founded during the oil crisis of 1973-74, the IEA's initial role was to co-ordinate measures in times of oil supply emergencies. As energy markets have changed, so has the IEA. Its mandate now incorporates work on climate change-policies, market reform, energy-technology collaboration and outreach to the rest of the world, especially major consumers and producers of energy including China, India, Russia and the OPEC countries.

Last November, the IEA released the 2008 edition of its *World Energy Outlook* (*WEO-2008*). The report concludes that it is not an exaggeration to claim that the future of human prosperity depends on how successfully we tackle the twin energy challenges facing us today: securing the supply of reliable and affordable energy; and effecting a rapid transformation to a low-carbon, efficient and environmentally benign system of energy supply. Current trends in energy supply and consumption point to rising imports of oil and gas into OECD regions and developing Asia while the growing concentration of production in an ever smaller number of countries threatens to increase our vulnerability to supply disruptions and sharp price hikes. And, in the absence of stronger policy action, rising consumption of fossil energy will drive up inexorably emissions and atmospheric concentrations of greenhouse gases, putting the world on track for an eventual global temperature increase of up to 6°C.

The report provides a more detailed assessment of oil-supply prospects than has ever before been released by the IEA. In a Reference Scenario, in which government policies are assumed to be unchanged, oil demand continues to grow strongly over the medium and longer-term. All of the projected increase is expected to come from non-OECD countries, led by China, India and the Middle East. The bulk of the increase in supply is expected to come from OPEC countries, their collective share rising from 41% today to 51% in 2030. Production has already peaked in most non-OPEC countries and will peak in most of the others before long. With respect to the United States, in the absence of a change in policy, we expect it to be importing around 12 mb/d of oil by 2030, only slightly down on current levels.

These trends point to a sea change in the structure of the upstream oil and gas industry. The international oil companies, which have traditionally dominated the sector, will be increasingly squeezed by the growing power of the national companies and by dwindling reserves and production in accessible mature basins outside OPEC countries. The challenges confronting the oil sector will be further exacerbated by the prospect of accelerating declines in production at individual oilfields. Based on the *WEO-2008's* detailed field-by-field analysis of the historical production trends of almost 800 of the world's oilfields – the most comprehensive study of its kind ever made public – we expect decline rates to accelerate significantly. Declines are fastest at oilfields in non-OPEC countries, including Mexico – a key supplier of crude oil to the United States.

Our analysis demonstrates that projections of oil supply are far more sensitive to assumptions about decline rates than to the rate of growth in oil demand. For instance, even if global oil demand was to remain flat until 2030, some 45 mb/d of additional gross capacity — the equivalent of over four times the current capacity of Saudi Arabia — would need to be brought on stream simply to offset declining production at existing fields.

The world's total endowment of oil is large enough to support the projected growth in output. The immediate risk to supply, however, is a lack of investment where it is needed. There remains a real possibility that under-investment will cause an oil-supply crunch in the medium term. More immediately, the credit crisis and deepening economic downturn is leading to a scaling back of all types of investment in most countries along the oil supply chain. While demand is also falling with the economic slump, there is a danger that investment in the coming months and years is reduced too much, leading to a shortage of capacity and another spike in prices several years later when the economy is on the road to recovery, due to the long lead times in completing large upstream and refining projects.

Although the trends that I have outlined are a cause for serious concern, they are not written in stone. Indeed there is much that can and is being done in many parts of the world to address the twin energy-related threats. In the past, the IEA has noted that very significant room remains to increase fuel-efficiency standards for trucks and cars in the United States, which would immediately contribute to energy and environmental security. In this respect, the new American Recovery and Reinvestment Act, with its strong focus on reducing fossil fuel dependence and greenhouse gas emissions by pursuing more aggressive demand-side and clean energy policies, is to be commended. Indeed we believe it makes good sense to exploit the opportunity of the financial and economic crisis to effect a shift in investment to low-carbon technologies. For example, the $95 billion that the IEA estimates the United States must invest each year in the power sector to move onto a pathway consistent with limiting the increase in the average global temperature to 2°C would also create jobs and enhance energy security.

Consideration could also now be given to taking advantage of the recent slide in the world oil price to review gasoline and diesel taxes and thereby "lock-in" the efficiency gains that resulted from last year's price surge. Similarly, I believe efforts to maximize the production of the United State's domestic oil and natural gas resource – including through an expansion of drilling on the Offshore Continental Shelf which is thought to contain significant amounts of recoverable resources – could form a crucial part of a comprehensive strategy to enhance the nation's energy security.

However, at the global level, the only possible solution to a long-term sustainable future is to strive for an energy mix that uses all options simultaneously. We need to combine greater energy efficiency improvements with more renewables and more nuclear. We must seek to minimise our dependence on fossil fuels while recognising that they will need to continue to make a significant contribution to meeting our energy needs for several decades to come: it is not realistic to expect low-carbon technologies to replace fossil energy overnight.

It is also imperative that international collaboration on energy policy is enhanced. Perhaps the best demonstration of this on the climate change front is that even if all OECD Member countries were to immediately reduce their CO_2 emissions to zero, we would still not be on a sustainable path unless non-OECD countries such as China, India and Russia were also to curb their emissions. IEA countries must also work with non-Members to address energy security, because all countries trade oil in an interconnected global market. Even if the United States were to succeed in lowering it oil imports in the coming years, increasing import dependency in other major consuming regions – notably China and India – would still mean that any oil supply disruption anywhere in the world would result in severe knock-on effects for the US market.

Mr. Chairman, and members of the Subcommittee, this completes my statement. I would be happy to take any questions you may have.

Chapter 2

ENERGY OUTLOOKS AND FEDERAL RESOURCES HEARING-GRUENSPECHT TESTIMONY

Howard Gruenspecht

Mr. Chairman and Members of the Committee, I appreciate the opportunity to appear before you today to discuss the U.S. energy outlook to 2030, focusing on the role of the Outer Continental Shelf (OCS) in current and projected energy production.

The Energy Information Administration (EIA) is the independent statistical and analytical agency within the Department of Energy that produces objective, timely, and relevant data, projections, and analyses to assist policymakers, help markets function efficiently, and inform the public. We do not promote, formulate, or take positions on policy issues, and our views should not be construed as representing those of the Department of Energy or the Administration.

THE ENERGY OUTLOOK: THE BIG PICTURE

The full *Annual Energy Outlook 2009* (*AEO2009*), which will be issued later this month, includes over 35 cases. The reference case and other *AEO2009* cases provide the results discussed in this testimony.

Liquid Fuels Consumption and Import Dependence. For the first time in more than 20 years, the *AEO2009* reference case projects no growth in U.S. oil consumption, reflecting the combined effect of recently enacted Corporate Average Fuel Economy standards, requirements for increased use of renewable fuels, and an assumed rebound in oil prices as the world economy recovers. With overall liquid fuel demand in the *AEO2009* reference case growing by only 1 million barrels per day between 2007 and 2030, plus increased use of domestically-produced biofuels and rising domestic oil production spurred by higher prices, the net import share of total liquids supplied, including biofuels, declines from 58 percent in 2007 to less than 40 percent in 2025 before increasing to 41 percent in 2030.

Natural Gas Consumption and Import Dependence. The reference case raises EIA's projection for U.S. production and consumption of natural gas compared to the previous *Annual Energy Outlook* (*AEO*), reflecting increased availability of resources and higher demand for electric power generation, due in part to the apparent impact of concerns related to greenhouse gas emissions on power plant investment decisions. With growing production of natural gas from unconventional onshore sources, the OCS, and Alaska, the net import share of total natural gas use also declines, from 16 percent in 2007 to less than 3 percent in 2030.

Total Primary Energy Use and Energy-Related Carbon Dioxide Emissions. Recently-enacted efficiency regulations and higher energy prices in the *AEO2009* reference case, compared to the last *AEO*, slow the rise in U.S. energy use, which is projected to grow from 101.9 quadrillion Btu in 2007 to 113.6 quadrillion Btu in 2030. When combined with the increased use of renewables and a reduction in projected additions of new coal-fired conventional power plants, this slows the growth in energy-related greenhouse gas emissions. Energy-related carbon dioxide emissions grow at 0.3 percent per year from 2007 to 2030 in the *AEO2009* reference case, reaching a level of 6,414 million metric tons in 2030, compared with 6,851 million metric tons in the *Annual Energy Outlook 2008* reference case.

Oil Prices. The assumption of a higher world oil price path in the *AEO2009* reference case reflects tighter constraints on access to low-cost oil supplies in a setting where the forces driving growth in long-term demand in countries outside of the Organization for Economic Cooperation and Development remain as strong as previously expected. The world crude oil price is projected to rise as the global economy rebounds and global demand once again grows more rapidly than non-Organization of Petroleum Exporting Countries liquids supply. In 2030, the average real price of crude oil is $130 per barrel in 2007 dollars ($189 per barrel in nominal dollars).

Renewable Energy Use. Total consumption of marketed renewable fuels—including wood, municipal waste, and biomass in the end-use sectors; hydroelectricity, geothermal, municipal waste, biomass, solar, and wind for electric power generation; ethanol for gasoline blending; and biomass-based diesel—grows by 3.3 percent per year in the *AEO2009* reference case. This rapid growth reflects the renewable fuel standard provisions included in the Energy Independence and Security Act of 2007 and strong growth in the use of renewables for electricity generation that is spurred by renewable portfolio standards for electricity generators in many States.

As requested by the Committee, the remainder of my testimony focuses more specifically on projections for oil and natural gas production from onshore and offshore resources, the factors that drive the projections, and sensitivity analyses under alternative access and price assumptions.

Table 1. Oil and Natural Gas Production from Federal Lands in Perspective, 2007

	Petroleum (million barrels)	Natural Gas (trillion cubic feet)
Production from Federal Lands	596	5.6
Onshore	105	2.8
Offshore	491	2.8
Other U.S. Production	1,253	13.5
Total U.S. Production	1,849	19.1
Total U.S. Consumption	7,548	23.0

Source: **Federal Onshore Production**: Minerals Management Service, Minerals Revenue Management, MRM WebStats, Federal Onshore Reported Royalty Revenues; **Total U.S. and Federal Offshore Oil Production and Total U.S. Petroleum Products Consumption**: Energy Information Administration (EIA), *Petroleum Supply Annual 2007*, DOE/EIA-0340(2007) (July 2008); **Total U.S. and Federal Offshore Natural Gas Production**: EIA, *Natural Gas Annual 2007*, DOE/EIA-013 1(2007) (January 2009); **U.S. Natural Gas Consumption**: EIA, *Annual Energy Review 2007*, DOE/EIA-0384(2007) (June 2008).

FEDERAL OFFSHORE AND ONSHORE RESOURCES IN CONTEXT

Resources on Federal lands, both offshore and onshore, are important to U.S. energy production. **Table 1** places onshore and offshore oil and natural gas production for 2007 in the context of total U.S. production and consumption. In 2007, roughly 32 percent of U.S. oil production and 29 percent of domestic natural gas production were from Federal lands.

Looking forward, production from Federal lands is expected to play an increasingly important role in total U.S. oil and natural gas production. Through 2030 the share of production from Federal lands is projected to increase to 47 percent for oil and 36 percent for natural gas (**Table 2**).

OCS PRODUCTION: HISTORICAL DATA AND NEAR-TERM FORECAST

OCS areas in the Western and Central portions of the Gulf of Mexico (GOM) are an important source of oil and natural gas production. In 2007, the GOM OCS areas, which have been producing substantial volumes of oil since the 1970s, produced 1.3 million barrels per day, amounting to about 25 percent of total U.S. crude oil production and down from peak OCS production of 1.6 million barrels per day in 2003. There are small amounts (less than 70 thousand barrels per day) of additional production from the Pacific OCS. Dry natural gas production in the GOM OCS in 2007 was 2.8 trillion cubic feet, down from peak production of 5.1 trillion cubic feet in 1997.

In the near term, OCS production is expected to rise as projects already under development come into operation. By 2012, projected GOM OCS oil production is 2.1 million barrels per day of oil and 3.4 trillion cubic feet of natural gas. As discussed below, forward-looking OCS production estimates to 2015 and later years, beyond the commissioning of projects already under development, are necessarily less certain since they are sensitive to the actual resource available, future prices, and future access to resources. However, using information from the Department of Interior's Minerals Management Service (MMS) regarding undiscovered technically recoverable resources, EIA data and MMS estimates regarding known reserves (proved reserves and projected reserve appreciation in known deposits), and assumptions regarding access policies, EIA develops projections of offshore oil and natural gas production through 2030.

Consistent with the *AEO* practice of reflecting existing laws and regulations, the *AEO2009* reference case reflects the removal in 2008 of the moratoria for drilling in the Atlantic, Pacific, and parts of the Eastern GOM OCS areas. Timing issues constrain the impacts of increased access in the near term. The MMS began the process of developing a leasing program that includes selected tracts from these areas after the moratoria were removed, with a timeline calling for the first leases to be offered in 2010. Once offered, leases must be bid on and awarded, and the wining bidders must develop and get approved exploration and development plans before any wells can be drilled. Thus, even if leasing were to begin next year, conversion of these newly-available resources to production would require some time. The *AEO2009* reference case assumes that the Pacific and Atlantic OCS regions are open for leasing starting in 2010 and that leasing begins in the Eastern GOM in 2022.

Based on the mean (50-percent probability) MMS estimate of undiscovered technically recoverable resources and estimates of known reserves and resources, the OCS areas that were until recently under moratoria in the Atlantic, Pacific, and Eastern GOM are estimated to hold about 20 percent of the total OCS technically recoverable oil resource (TROR)—1 8 billion barrels out of a total of more than 93 billion barrels, exclusive of past production as of January 1, 2007. The estimates of TROR in the GOM OCS areas open to leasing prior to 2008 and the Alaska OCS are 47 billion barrels and 27 billion barrels, respectively. According to MMS estimates, there is only a 5-percent chance that OCS areas formerly under moratoria have more than 27 billion barrels of TROR.

Based on the MMS mean estimate of undiscovered technically recoverable natural gas resources and estimates of known reserves and resources, total technically recoverable natural gas resources in the OCS are estimated at 456 trillion cubic feet as of January 1, 2007. Roughly 76 trillion cubic feet (or 17 percent) are estimated to be in areas formerly under moratoria in the Atlantic, Pacific and Eastern GOM—nearly half or 37 trillion cubic feet in the Atlantic, 18 trillion cubic feet in the Pacific, and 21 trillion cubic feet in the Eastern GOM.

Table 2. Projected Oil and Natural Gas Production on Federal Lands Compared to Projected U.S. Total Production

Year	Crude Oil (million barrels)			Natural Gas (trillion cubic feet)		
	Offshore Federal	Onshore[a] Federal	Total U.S.	Offshore Federal	Onshore[a] Federal	Total U.S
2008	468	116	1,808	2.9	3.0	20.5
2010	714	118	2,051	3.0	3.0	20.4
2025	953	228	2,633	4.9	3.5	23.2
2030	986	276	2,690	4.7	3.8	23.6

[a] Federal onshore production is not explicitly represented in the National Energy Modeling System. The volumes are estimated based on historical trends and the projected regional production from the reference case of the *Annual Energy Outlook 2009*.

Source: Energy Information Administration, *Annual Energy Outlook 2009*, DOE/EIA-0383(2009).

Assumptions about exploration, development, and production of economical fields, such as drilling schedules, costs, platform selection, reserves-to-production ratios, etc., in the Pacific, Atlantic, and Eastern GOM are generally based on data for fields in the Central GOM that are of similar water depth and size. In addition, it is assumed that local infrastructure issues and other potential non-Federal impediments are resolved. Lack of resolution of these issues would, of course, affect the projections.

Lower-48 offshore crude oil production is projected to increase from 1.4 million barrels per day in 2007 to 2.7 million barrels per day in 2030. Production from new OCS leases in the Pacific is projected to begin in 2015, with total Pacific production reaching nearly 0.5 million barrels per day in 2030. Crude oil production from the Atlantic region is projected to begin in 2019, reaching 0.2 million barrels per day by 2030. Crude oil production in all areas of the GOM rises from 1.3 million barrels per day to 2.1 million barrels per day between 2007 and 2030.

Estimates of production from the OCS areas previously under moratoria are higher than in a previous analysis presented in the *Annual Energy Outlook 2007* primarily because the *AEO2009* has significantly higher oil and natural gas prices and because the assumed initial flow rate of Pacific OCS fields in shallow waters was adjusted to better reflect the production potential from these oil-prone fields compared to more natural-gas-prone fields in similar water depth and size in the Central GOM.

Lower-48 offshore natural gas production is projected to increase from 3.0 trillion cubic feet in 2007 to 4.9 trillion cubic feet in 2030. By 2030, Pacific natural gas production is projected to reach nearly 0.3 trillion cubic feet and production from the Atlantic region is projected to reach 0.5 trillion cubic feet.

EIA's OCS Estimates: Discussion and Comparison with Historical Experience

One way to gain perspective on EIA's estimates of production in OCS areas formerly under moratoria is to consider how the relationship between projected production and MMS indicators of resource levels and characteristics in those areas compares to that for the GOM OCS area that was open prior to 2008.

TROR Comparisons. Oil reserves in the GOM OCS area open before 2008, which has already been leased and developed extensively, are about 4 billion barrels, with an additional 9 billion barrels of expected reserve appreciation in discovered fields. Adding the estimate of 34 billion barrels of undiscovered TROR, the mean estimate of total TROR in the GOM area open before 2008 is 47 billion barrels, which is more than 2.5 times the MMS mean estimate of 18 billion barrels of TROR in OCS areas formerly under moratoria.

Average Field Size Comparisons. Field size matters because larger fields are more attractive development targets than smaller ones. The average size across all existing GOM OCS oil and natural gas fields is 43 million barrels of oil equivalent. MMS has also developed field size distributions for undiscovered OCS fields that it used to prepare reports mandated under the Energy Policy Act of 2005. The MMS estimate of the average undiscovered field size in GOM OCS areas open to drilling prior to 2008 is 59 million barrels of oil equivalent, which is significantly greater than the average field size of 15 million barrels of oil equivalent for OCS areas formerly under moratoria.

Other Project Development Factors. Project development time frames and expected returns vary substantially across offshore projects depending upon such factors as: 1) size of the field; 2) relative proportion of oil, natural gas, and condensates in the field; 3) reservoir and oil characteristics,;4) water depth; 5) distance to nearest oil and/or natural gas pipelines; 6) whether there are other nearby fields to share in the expense of building new pipelines; and 7) the type of production system chosen for field development, e.g., anchored platform, tension-leg platform, tethered spar, or floating production storage, and offloading ship.

To the extent that information is available, the indicators of resource levels and characteristics for the OCS areas previously under moratoria are generally inferior to those for the GOM OCS open prior to 2008, as discussed above. This is reflected in EIA's view that, through 2030, access to the OCS areas formerly under moratoria adds only a fraction of the daily production volume provided by the GOM OCS area open prior to 2008.

EIA recognizes that all forward-looking production estimates are inherently uncertain. Some factors that could lead to higher daily production estimates for the OCS areas formerly under moratoria include the use of the 5-percent, or 1-in-20, probability estimate of TROR and the assumption of a more favorable field size distribution than that used by MMS in its recently published reports. Consideration of any long-term constraints on rig availability that reflect the prioritization of alternative offshore projects or the possibility that non-Federal impediments to production would persist over time could result in lower daily production estimates.

AEO2009 ACCESS SENSITIVITY CASE

As part of the *AEO2009*, EIA prepared a restored moratoria sensitivity case. U.S. OCS crude oil production in 2030 is projected to be 565,000 barrels per day lower in the restored moratoria case than in the reference case—2.2 million barrels per day compared to 2.7 million barrels per day. Cumulative domestic production of crude oil from both onshore and offshore sources between 2010 and 2030 in the restored moratoria case is projected to be 2.1 billion barrels, or 4.2 percent, lower than in the *AEO2009* reference case.

As with oil, access to OCS resources affects the domestic supply of natural gas. However, because the volume of technically recoverable natural gas in the OCS areas previously under moratoria accounts for less than 5 percent of the total U.S. technically recoverable natural gas resource base, the volume impacts are smaller relative to the baseline supply level. Cumulatively, domestic natural gas production from both onshore and offshore sources between 2010 through 2030 is projected to be 1.3 percent lower in the restored moratoria case than in the *AEO2009* reference case. Natural gas production from the Lower-48 offshore in 2030 is projected to be 4.1 trillion cubic feet in the restored moratoria case compared to 4.9 trillion cubic feet in the *AEO2009* reference case. In contrast to the situation in oil, the reduction in offshore supply of natural gas in the restored moratoria case is partially offset by an increase in onshore production. Reduced OCS access in the restored moratoria case results in higher natural gas prices, which increase the projection for U.S. onshore gas production by 0.2 trillion cubic feet in 2030 compared to the level in the reference case.

AEO2009 LOW PRICE SENSITIVITY CASE

The impact of access to OCS resources on domestic production is lessened in the low price case, where oil prices are assumed to remain near $50 per barrel (2007 dollars) through 2030, rather than rising to $110 per barrel by 2015 and $130 per barrel (2007 dollars) by 2030 as assumed in the reference case. In 2030, total OCS crude oil production is projected to be 440,000 barrels per day higher in the low world oil price case than in the low oil price case with the OCS moratoria reinstated—2.1million barrels per day compared with 1.7 million barrels per day. The observation that U.S. OCS production in 2030 under reference case prices with full restoration of the OCS moratoria, at 2.2 million barrels per day, is projected to be higher than U.S. OCS production in the low price case with no moratoria underlines the importance of prices as a determinant of future production.

The OCS is expected to remain a major contributor to domestic crude oil and natural gas supply under a variety of price and access assumptions. Although a significant volume of undiscovered technically recoverable oil and natural gas resources has been added with access to the Pacific, Atlantic, and parts of the Eastern GOM OCS, there is a great deal of uncertainty surrounding the resource estimates as well as the timing and cost to explore and develop these resources.

This concludes my statement, Mr. Chairman. I will be happy to answer any questions you and the other Members may have.

In: U.S. Energy: Overview of the Trends... ISBN: 978-1-60876-041-1
Editors: Gregor E. Peake © 2010 Nova Science Publishers, Inc.

Chapter 3

ENERGY OUTLOOKS AND FEDERAL RESOURCES HEARING-PIERCE TESTIMONY

Brenda S. Pierce

Mr. Chairman and Members of the Subcommittee, thank you for the opportunity to appear here today to discuss with you the U.S. Geological Survey's role in studying, understanding, and assessing the undiscovered, geologically based, energy resources of the Nation (exclusive of the Federal offshore) and World and the Minerals Management Services' (MMS) role in providing information on Federal resources of the Outer Continental Shelf (OCS).

INTRODUCTION

Adequate, reliable, and affordable energy supplies obtained using environmentally sustainable practices are essential to economic prosperity, environmental and human health, and political stability. National and global energy demand and resource consumption are projected to increase over the next several decades, though at a slower rate than in recent years. The United States currently consumes 21 percent of the total world primary energy consumption and produces 15 percent of the total world primary energy production. Thus, the volumes, quality, and availability of domestic and foreign energy resources are of critical importance to the United States. The Nation continues to face important decisions regarding the competing uses of public lands and offshore waters, the supply of energy to sustain development and enable growth, and the environmental effects of energy resource development.

ROLE OF THE U.S. GEOLOGICAL SURVEY

The U.S. Geological Survey (USGS) provides the research and information needed to address these challenges by conducting scientific investigations of geologically based energy resources, such as research and assessment on the geology of oil, gas, and coal resources, emerging resources such as gas hydrates, underutilized resources such as geothermal, and unconventional resources such as oil shale, and research on the effects associated with energy resource occurrence, production, and (or) utilization. Our goal is: (1) to understand the processes critical to the formation, accumulation, occurrence, and alteration of geologically based energy resources; (2) to conduct scientifically robust assessments of those resources; and (3) to study the impact of energy resource occurrence and (or) production and use on both environmental and human health. The results from these geoscientific studies are used to evaluate the quality and distribution of energy resource accumulations, and to assess the energy resource potential of the Nation (exclusive of the Federal OCS)) and the World. As one example, the USGS recently produced the first-ever estimate of undiscovered, technically recoverable gas from natural gas hydrates. Although these resources have not yet been proven economic, this USGS assessment estimates a mean of 85.4 trillion cubic feet of technically recoverable gas from gas hydrates on the Alaska North Slope.

The results from this and other USGS research provide impartial, robust scientific information about energy resources that directly supports the U.S. Department of the Interior's mission of protecting and responsibly managing the Nation's natural resources. The USGS and MMS information is used by policy and decision makers, land and resource managers, other Federal and State agencies, the domestic energy industry, foreign governments, nongovernmental groups, academia, other scientists, and the public. The USGS works with the MMS, which has responsibility for energy and minerals management in Federal offshore waters, to provide an integrated evaluation of the Nation as a whole. Collectively, information from USGS research advances the scientific understanding of energy resources, contributes to plans for a balanced and secure energy future, and facilitates the strategic use and evaluation of resources.

USGS AND MMS NATIONAL OIL AND GAS RESOURCES RESEARCH AND ASSESSMENT ACTIVITIES

The overall goal of USGS domestic energy activities is to conduct research and assessments of all geologically based energy resources. This includes undiscovered, technically recoverable oil and natural gas resources, both conventional and unconventional of the United States (exclusive of the Federal OCS, which is assessed by the MMS). These are resources that have yet to be found (drilled), but if found, could be recovered using currently available technology and industry practice. Economic factors are not always considered; for example, it may not be economically feasible to exploit gas hydrate resources on the Alaska North Slope and both conventional and unconventional Alaskan gas resources are currently considered stranded without the means of transporting gas from the region. The purpose of USGS and MMS assessments are to develop robust, geologically based, statistically sound, well-documented estimates of quantities of energy resources having the potential to be added to reserves, and thus contribute to the overall energy supply. The USGS and MMS resource assessment methodologies are thoroughly reviewed and externally vetted so as to maintain the transparency and robustness of the assessment results.

The current USGS effort to update national (onshore and State waters) assessments of oil and gas resources is done in support of the Energy Policy and Conservation Act (EPCA) Amendments of 2000 (P.L. 106–469 §604). Through a collaborative, multi-agency effort involving the Bureau of Land Management, the USGS, the U.S. Forest Service, the Department of Energy, and the EIA, the USGS provides the oil and gas resource estimates as the basis for the EPCA inventory. The USGS role is to assess the potential volumes of conventional and continuous (unconventional) resources (e.g., coalbed gas, shale gas, tight gas sands) in each priority province using established, externally reviewed and vetted methodologies and provide this information to the appropriate land and resource management agencies for subsequent analysis. The Energy Policy Act of 2005 (P.L. 109-58) re-authorized EPCA 2000 assessment activities by the USGS, emphasizing the unique and critical role of the USGS and specifically mandated that "the same assessment methodology across all geological provinces, areas, and regions [be used] in preparing and issuing national geological assessments to ensure accurate comparisons of geological resources."

The estimate of undiscovered, technically recoverable resources changes over time. There are several reasons for this, including scientific and technological developments regarding petroleum resources in general and improvements to the geologic understanding in numerous settings. These advances in geologic understanding, as well as changes in technology and industry practices, necessitate that resource assessments be periodically updated to take into account such advances. One example of this change is the recently updated USGS assessment of the Bakken Formation in the U.S. portion of the Williston Basin. This assessment, released in 2008, shows an estimated 3.0 to 4.3 billion barrels of undiscovered, technically recoverable oil compared to USGS's 1995 mean estimate of 151 million barrels of oil. Another example is the USGS assessment of gas hydrates on the Alaskan North Slope. Substantial investments in gas hydrate research now support categorizing some accumulations of gas hydrates as technically recoverable. Research challenges remain in order to determine if this technically recoverable resource will be economically recoverable, but current multi-organizational (including USGS) and multi-disciplinary efforts are focused on overcoming these obstacles.

The passage of the OCS Lands Act in 1953 established Federal jurisdiction over the mineral resources of the OCS and authorized the Secretary of the Interior to manage oil and natural gas and other marine minerals activity seaward of state submerged lands. Oil and natural gas produced offshore on the OCS is a major supply source of energy for the domestic market. About 17 billion barrels of oil and 174 trillion cubic feet of natural gas have been produced from the OCS since 1954. Current production levels are about 1.4 million barrels of oil and about 8 billion cubic feet of natural gas per day. This represents approximately 27 percent of domestic oil production and 14 percent of natural gas production. But these shares are expected to grow over the next 7 years as new deepwater production in the Gulf of Mexico comes on line *(Gulf of Mexico Oil and Gas Production Forecast: 2007-2016, May 2007)*. Recent discoveries in the deep and ultra-deep waters of the Gulf of Mexico could help provide a significant source of oil and gas supplies for decades to come.

OCS oil and gas resource assessments are completed as part of the Secretary's responsibilities for managing OCS energy and mineral resources and the requirement to assure fair market value for OCS lands to be leased. The MMS conducts resource assessments for the OCS at various scales and for many purposes. Regional assessments may be prepared simply to develop an inventory of potential oil and natural gas resources as part of an evaluation of future supply options. Assessments may be undertaken to analyze the relative merits of oil and gas development proposals and alternatives versus other competing uses. Resource estimates also provide critical input to decision makers regarding the virtues of various policy alternatives, and provide data essential for valuing Federal lands prior to leasing or analyzing industry exploration or development proposals. The MMS conducts periodic national assessments of the oil and natural gas resource potential of the Nation's Outer Continental Shelf; and in 2005, Congress directed (in Section 357 of the Energy Policy Act of 2005) that the Secretary conduct a comprehensive inventory and analysis of oil and natural gas resources of the U.S. OCS. This MMS assessment, which was completed in 2006, considers recent geophysical, geological, technological, and economic information and utilizes a probabilistic play based approach to estimate the undiscovered technically recoverable resources (UTRR) of oil and gas for individual plays. This methodology is suitable for both conceptual plays where there is little or no specific information available, and for developed plays where there are discovered oil and gas fields and considerable information is available. After estimation, individual play results are aggregated to larger areas such as basins and regions. Estimates of the quantities of historical production, reserves, and future reserves appreciation are presented to provide a frame of reference for analyzing the estimates of UTRR.

Reserve growth is well documented in the United States and is a major component of the Nation's remaining oil and natural gas resources. In fact, most additions to world oil reserves in recent years are from growth of reserves in existing fields rather than new discoveries. The EIA's 2009 forecast of significant increases in domestic oil production is partly owing to advances in enhanced oil recovery technologies. Given this context, it is important to note the important distinction between the terms "resource" and "reserves." Resource is a concentration of naturally occurring solid, liquid, or gaseous hydrocarbons in or on the Earth's crust, some of which is, or potentially is, economically extractable. Reserves specifically refer to the estimated quantities of identified (discovered) petroleum resources that as of a specified date, are expected to be commercially recovered from known accumulations under prevailing economic conditions, operating practices, and government regulations.

Reserve growth occurs for a variety of reasons, including: (1) extensions of existing fields, infill drilling and new pool discoveries, (2) application of new recovery technologies and improved efficiency, and (3) revisions resulting from recalculation of viable reserves in dynamically changing economic and operating conditions. The assessment of both undiscovered resources and of additions to reserves from discovered fields and reservoirs requires estimation of reserve growth. The USGS has an active research effort to develop a methodology and approach for better quantifying domestic and global contributions of reserve growth to the petroleum resource endowment.

Undiscovered, technically recoverable mean oil resources total 48 billion barrels of oil onshore and in State waters and 86 billion barrels of oil for the OCS. Undiscovered, technically recoverable mean natural gas resources total 743 trillion cubic feet onshore and in State waters (or 657 trillion cubic feet, exclusive of the recent natural gas hydrates assessment), and 420 trillion cubic feet for the OCS. These resources have the potential to be added to reserves, but are not yet proven and may or may not be economic at current or future prices. For example, according to the 2006 MMS national assessment (*http://www.mms.gov/revaldiv/PDFs/NA2006BrochurePlanningAreaInsert. pdf*), of the 86 billion barrels of undiscovered, technically recoverable oil resources in the OCS, 54 billion barrels of that is estimated to be economically recoverable at $46/barrel. Of the 420 trillion cubic feet of undiscovered, technically recoverable natural gas resources in the OCS, 215 trillion cubic feet is estimated to be economically recoverable at $6.96/million cubic foot."

These numbers can be compared to proved reserves numbers (EIA): proved U.S. petroleum reserves (for 2007) are 22 billion barrels of oil and proved world petroleum reserves are 1,317 billion barrels; proved natural gas reserves for the U.S. are 204 trillion cubic feet and for the world are 6,124 trillion cubic feet.

UNCONVENTIONAL OIL AND GAS RESOURCES

In April 2007, the USGS received funding for a two-year project to reassess oil shale deposits of the Eocene Green River Formation of Colorado, Utah, and Wyoming. The new assessment will incorporate considerable data acquired by the USGS following the collapse of the oil shale industry in the

1980's. It will subdivide the oil shale section into various subunits that will be assessed separately and the data will be made available on-line in a manner that can be easily utilized by modern computer models. This will allow simulations of various development scenarios for open pit mining, underground mining, and in-situ retorting, should oil shale development ever get underway.

COAL

Coal dominates the U.S. fossil energy endowment and accounts for 48% of domestic electricity generation. The USGS has recently completed an assessment of coal resources and reserves in Wyoming's Gillette coalfield, the most prolific coalfield in the country. This assessment is part of the National Coal Resource and Reserve Assessment, which is systematically evaluating the domestic coal resource and reserve base. By utilizing an abundance of new data from coalbed methane development in the region, the USGS was able to produce the most comprehensive assessment to date. The Gillette area accounts for nearly 40 percent of the Nation's current coal production making it the single most important coalfield in the United States. A total of 164 billion tons of original coal resources was found in the six coal beds included in the evaluation. Of that original resource, 10.1 billion tons (6 percent) can be classified as reserves at the current average estimated sales price. Substantial additional resources could be recoverable assuming increased market prices will support the higher costs needed to recover deeper coal. Coal is currently the most important fuel for electricity generation and the USGS studies will determine what portion of the resource base is technically and economically recoverable.

RENEWABLE ENERGY

In addition to petroleum and coal resources, the USGS also evaluates renewable resources such as geothermal energy. The USGS recently completed a national geothermal resource assessment, the first one in more than 30 years. The USGS evaluated 241 moderate- and high-temperature geothermal resources capable of producing electricity. The USGS assessment estimates (1) 9,057 Megawatts-electric (MWe) of power potential from conventional, identified geothermal systems, (2) 30,033 MWe of power generation potential from conventional, undiscovered geothermal resources, and (3) 517,800 MWe of power generation potential from unconventional Enhanced Geothermal Systems (EGS) resources. The results indicate that full development of the conventional, identified systems could expand geothermal power production by approximately 6,500 MWe, or about 260 percent of the currently installed geothermal total of more than 2,500 MWe. The resource estimate for unconventional EGS is more than an order of magnitude larger than the combined estimates of both identified and undiscovered conventional geothermal resources and, if successfully developed, could provide an installed geothermal electric power generation capacity equivalent to about half of the currently installed electric power generating capacity of the United States.

America's oceans may also provide potential new renewable energy sources to support our Nation's growing energy needs, and MMS is developing a program for managing their uses. Resources on the OCS can be used to generate electricity in a variety of ways. To date there is no comprehensive evaluation for the available renewable energy potential in all offshore waters, but researchers have begun to examine the resource potential in specific areas of interest. DOE's National Renewable Energy Laboratory has a program to produce validated wind resource maps for priority offshore areas, and the results show that the offshore wind resource potential is vast and has the potential to meet a significant amount of the Nation's future energy needs. Although significant wind, wave, tidal and current resources exist in close proximity to coastal population centers—areas that consume the majority of the Nation's electricity generation—the technologies used to generate this energy are relatively new and untested in the offshore environment of the U.S. OCS. Wind, wave and ocean current technologies have been demonstrated at the pilot scale, and wind has been developed at the commercial scale outside the United States—e.g., offshore Denmark, the United Kingdom and Germany.

U.S. GEOLOGICAL SURVEY INTERNATIONAL ENERGY STUDIES

Our Nation depends heavily on imported energy resources: about 58 percent of the oil and 16 percent of the natural gas consumed in the US come from imports. Given the significance of imported oil and gas to the U.S. energy mix, scientifically robust, unbiased assessments of the world's remaining endowment of petroleum accumulations are of the utmost importance. For this reason, global petroleum resource assessments are a core USGS research activity and have significant global visibility. The USGS world oil and gas resource estimates are used as a standard reference by many organizations including the EIA and the International Energy Agency (IEA).

The overall objectives of USGS studies of international petroleum resources are to continue providing high-quality, comprehensive petroleum assessments and to update previous assessments as needed. A major focus of recent USGS research in this area is the Circum-Arctic Resource Appraisal (or CARA), the primary emphasis of which is to provide a comprehensive, unbiased probabilistic estimate of potential future additions to conventional oil and gas reserves in the high northern latitudes. The Arctic is an area of high petroleum resource potential, low data density, high geologic uncertainty and sensitive environmental conditions. The assessment is the first publicly available petroleum resource estimate of the entire area north of the Arctic Circle.

The results of the assessment, released last July, estimate that the area north of the Arctic Circle has 90 billion barrels of undiscovered, technically recoverable oil, 1,670 trillion cubic feet of technically recoverable natural gas, and 44 billion barrels of technically recoverable natural gas liquids in 25 geologically defined areas thought to have potential for petroleum. These resources account for about 22 percent of the undiscovered, technically recoverable resources in the world. The Arctic accounts for about 13 percent of the undiscovered oil, 30 percent of the undiscovered natural gas, and 20 percent of the undiscovered natural gas liquids in the world. About 84 percent of the estimated resources are expected to occur offshore.

CONCLUSION

During the next decade, the Federal government, industry, and other groups will need to better understand the domestic and global distribution of, genesis of, use of and consequences of using geologically based energy resources to address pressing environmental problems such as climate change, national security issues, manage the Nation's domestic supplies wisely, predict future needs, anticipate as well as guide changing patterns in use, and facilitate creation of new industries. Energy resources research and assessments are a traditional strength of the USGS and the MMS, and these activities provide impartial, robust information necessary for the many needs just outlined. As the Nation's energy mix evolves, the USGS and MMS will continue to work with other Federal agencies such as DOE to ensure that our research and assessment portfolio ties into a comprehensive suite of assessments to inform policymakers about energy choices. Future USGS and MMS assessments are anticipated to include hydrocarbon-based (for example, unconventional gas from coal and shale, gas hydrates, oil shale) and nonhydrocarbon-based sources (for example, geothermal resources and uranium) and address the effects of such resource use on land use, ecosystem health, and human welfare. USGS resource assessments and research play an important role in the public and government discourse about the energy resource future of the Nation so that science can inform, advise, and engage decision makers. The USGS and MMS stand ready to assist Congress as it examines these challenges and opportunities.

Thank you for this opportunity to provide an overview of USGS and MMS research and assessments of geologically based energy resources. I would be happy to answer your questions.

In: U.S. Energy: Overview of the Trends... ISBN: 978-1-60876-041-1
Editors: Gregor E. Peake © 2010 Nova Science Publishers, Inc.

Chapter 4

U.S. ENERGY: OVERVIEW AND SELECTED FACTS AND NUMBERS

Carl E. Behrens and Carol Glover

SUMMARY

Energy supplies and prices are major economic factors in the United States, and energy markets are volatile and unpredictable. Thus, energy policy has been a recurring issue for Congress since the first major crisis in the 1970s. As an aid in policy making, this report presents a current and historical view of the supply and consumption of various forms of energy.

The historical trends show petroleum as the major source of energy, rising from about 38% in 1950 to 45% in 1975, then declining to about 40% in response to the energy crisis of the 1970s. Significantly, the transportation sector has been and continues to be almost completely dependent on petroleum, mostly gasoline. The importance of this dependence on the volatile world oil market was revealed over the past five years as perceptions of impending inability of the industry to meet increasing world demand led to relentless increases in the prices of oil and gasoline. With the downturn in the world economy and a consequent decline in consumption, prices collapsed, but the dependence on imported oil continues as a potential problem.

Natural gas followed a similar pattern at a lower level, increasing its share of total energy from about 17% in 1950 to more than 30% in 1970, then declining to about 20%. Consumption of coal in 1950 was 35% of the total, almost equal to oil, but it declined to about 20% a decade later and has remained at about that proportion since then. Coal currently is used almost exclusively for electric power generation.

Nuclear power started coming online in significant amounts in the late 1960s. By 1975, in the midst of the oil crisis, it was supplying 9% of total electricity generation. However, increases in capital costs, construction delays, and public opposition to nuclear power following the Three Mile Island accident in 1979 curtailed expansion of the technology, and many construction projects were cancelled. Continuation of some construction increased the nuclear share of generation to 20% in 1990, where it remains currently. The first new reactor license applications in nearly 30 years were recently submitted, but no new plants are currently under construction or on order.

Construction of major hydroelectric projects has also essentially ceased, and hydropower's share of electricity generation has gradually declined, from 30% in 1950 to 15% in 1975 and less than 10% in 2000. However, hydropower remains highly important on a regional basis.

Renewable energy sources (except hydropower) continue to offer more potential than actual energy production, although fuel ethanol has become a significant factor in transportation fuel, and wind power has recently grown rapidly. Conservation and energy efficiency have shown significant gains over the past three decades and offer encouraging potential to relieve some of the dependence on imports that has caused economic difficulties in the past, as well as the present.

After an introductory overview of aggregate energy consumption, this report presents detailed analysis of trends and statistics regarding specific energy sources: oil, electricity, natural gas, coal and renewable energy. A section on trends in energy efficiency is also presented.

INTRODUCTION

Tracking changes in energy activity is complicated by variations in different energy markets. These markets, for the most part, operate independently, although events in one may influence trends in another. For instance, oil price movement can affect the price of natural gas, which then plays a significant role in the price of electricity. Since aggregate indicators of total energy production and consumption do not adequately reflect these complexities, this compendium focuses on the details of individual energy sectors. Primary among these are oil, particularly gasoline for transportation, and electricity generation and consumption. Natural gas is also an important energy source, for home heating as well as in industry and electricity generation. Coal is used almost entirely for electricity generation, nuclear and hydropower completely so.

Renewable sources (except hydropower) continue to offer more potential than actual energy production, although fuel ethanol has become a significant factor in transportation fuel, and wind power has recently grown rapidly. Conservation and energy efficiency have shown significant gains over the past three decades, and offer encouraging potential to relieve some of the dependence on imports that has caused economic difficulties in the past as well as the present.

To give a general view of energy consumption trends, **Table 1** shows consumption by economic sector—residential, commercial, transportation, and industry—from 1950 to the present. To supplement this overview, some of the trends are highlighted by graphs in **Figures 1 and 2**.

In viewing these figures, a note on units of energy may be helpful. Each source has its own unit of energy. Oil consumption, for instance, is measured in million barrels per day (mbd),[1] coal in million tons per year, natural gas in trillion cubic feet (tcf) per year. To aggregate various types of energy in a single table, a common measure, British Thermal Unit (Btu), is often used. In **Table 1**, energy consumption by sector is given in units of quadrillion Btus per year, or "quads," while per capita consumption is given in million Btus (MMBtu) per year. One quad corresponds to one tcf of natural gas, or approximately 50 million tons of coal. One million barrels per day of oil is approximately 2 quads per year. One million Btus is equivalent to approximately 293 kilowatt-hours (Kwh) of electricity. Electric power generating capacity is expressed in terms of kilowatts (Kw), megawatts (Mw, equals 1,000 Kw) or gigawatts (Gw, equals 1,000 Mw). Gas-fired plants are typically about 250 Mw, coal-fired plants usually more than 500 Mw, and large nuclear powerplants are typically about 1.2 Gw in capacity.

Table 1 shows that total U.S. energy consumption almost tripled since 1950, with the industrial sector, the heaviest energy user, growing at the slowest rate. The growth in energy consumption per capita (i.e., per person) over the same period was about 50%. As **Figure 1** illustrates, much of the growth in per capita energy consumption took place before 1970.

Table 1 does not list the consumption of energy by the electricity sector separately because it is both a producer and a consumer of energy. For the residential, commercial, industrial, and transportation sectors, the consumption figures given are the sum of the resources (such as oil and gas) that are directly consumed plus the total energy used to produce the electricity each sector consumed—that is, both the energy value of the kilowatt-hours consumed and the energy lost in generating that electricity. As **Figure 2** demonstrates, a major trend during the period was the electrification of the residential and commercial sectors and, to a lesser extent, industry. By 2007, electricity (including the energy lost in generating it) represented about 70% of residential energy consumption, about 80% of commercial energy consumption, and about a third of industrial energy consumption.[2]

Table 1. U.S. Energy Consumption, 1950-2007

	Energy Consumption by Sector (Quadrillion Btu)					Population (millions)	Consumption Per Capita (Million Btu)		
	Resid.	Comm.	Indust.	Trans.	Total		Total	Resid.	Trans.
1950	6.0	3.9	16.2	8.5	34.6	152.3	227.3	39.4	55.8
1955	7.3	3.9	19.5	9.6	40.2	165.9	242.3	44.0	57.6
1960	9.1	4.6	20.8	10.6	45.1	80.7	249.6	50.2	58.7
1965	10.7	5.8	25.1	12.4	54.0	94.3	278.0	55.0	64.0
1970	13.8	8.3	29.6	16.1	67.8	205.1	330.9	67.3	78.5
1975	14.8	9.5	29.4	18.2	72.0	216.0	333.4	68.7	84.5
1980	15.8	10.6	32.1	19.7	78.1	227.2	343.8	69.5	86.7
1985	16.1	11.4	28.9	20.1	76.5	237.9	321.5	67.6	84.4
1990	17.0	13.3	31.9	22.4	84.7	249.6	339.1	68.2	89.8
1995	18.6	14.7	34.0	23.8	91.2	266.3	342.4	69.8	89.6
2000	20.5	17.2	34.8	26.6	99.0	282.2	350.7	72.6	94.1
2005	21.7	17.9	32.5	28.4	100.5	295.9	339.7	73.4	95.8
2006	20.9	17.7	32.4	28.9	99.9	298.8	334.2	69.8	96.6
2007P	21.8	18.4	32.3	29.1	101.6	301.6	336.8	72.1	96.5

Source: Energy Information Administration (EIA), Annual Energy Review 2007, Tables 2.1a and D1. Per capita data calculated by CRS.
Notes: Data for 2007 are preliminary.

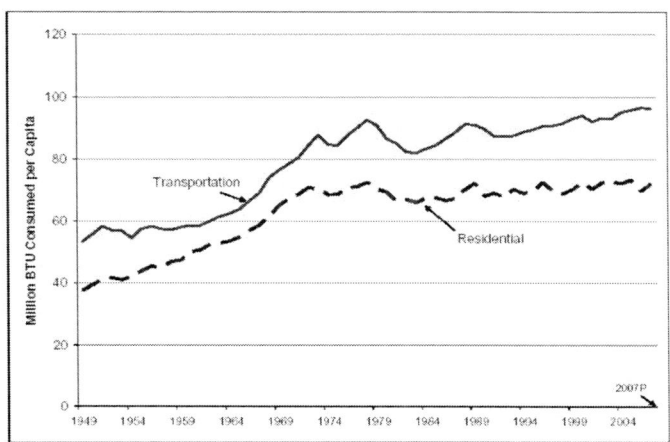

Source: Energy Information Administration (EIA), Annual Energy Review 2007, Tables 2.1a and D1. Per capita data calculated by CRS.
Notes: Data for 2007 are preliminary.

Figure 1. Per Capita Energy Consumption in Transportation and Residential Sectors, 1949-2007

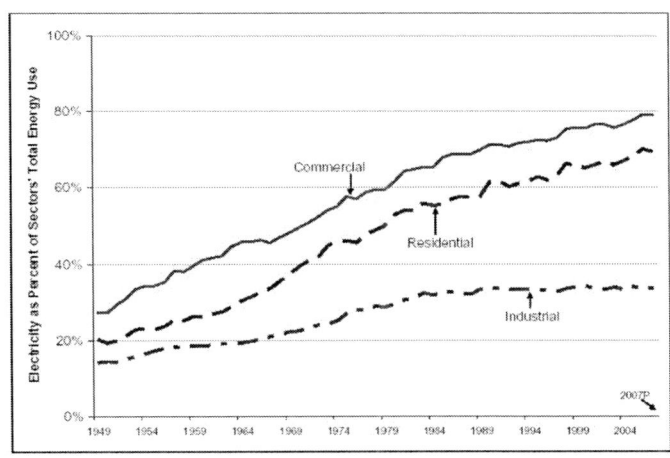

Source: Energy Information Administration (EIA), Annual Energy Review 2007, Tables 2.1a and D1. Per capita data calculated by CRS.
Notes: Data for 2007 are preliminary.

Figure 2. Electricity Intensity: Commercial, Residential, and Industrial Sectors, 1949-2007

Consumption of major energy resources—petroleum, natural gas, and coal—is presented in **Table 2** and **Figure 3**. The historical trends show that petroleum has been and continues to be the major source of energy, rising from about 38% in 1950 to 45% in 1975, then declining to about 40% in response to the energy crisis of the 1970s. Natural gas followed a similar pattern at a lower level, increasing its share of total energy from about 17% in 1950 to over 30% in 1970, then declining to about 20%. Consumption of coal in 1950 was 35% of the total, almost equal to oil, but it declined to about 20% a decade later and has remained at about that proportion since then.

Table 2. Energy Consumption in BritishThermal Units (BTU) and as a Percentage of Total, 1950-2007 (Quadrillion BTU)

	Petroleum		Natural Gas		Coal		Other		Total
	Quads	% of total	Quads	% of total	Quads	% of total	Quads	% of total	
1950	13.3	38.4	6	17.3	12.3	35.5	3.0	8.6	34.6
1955	17.3	43.0	9	22.4	11.2	27.8	2.8	7.0	40.2
1960	19.9	44.1	12.4	27.5	9.8	21.7	3.0	6.5	45.1
1965	23.2	42.9	15.8	29.2	11.6	21.4	3.4	6.4	54.0
1970	29.5	43.5	21.8	32.1	12.2	18.0	4.3	6.4	67.8
1975	32.7	45.4	19.9	27.6	12.7	17.7	6.6	9.2	72.0
1980	34.2	43.8	20.4	26.1	15.4	19.7	8.3	10.6	78.1
1985	30.9	40.4	17.8	23.3	17.5	22.9	10.4	13.6	76.5
1990	33.6	39.7	19.7	23.3	19.2	22.7	12.3	14.6	84.7
1995	34.6	37.9	22.8	25.0	20.2	22.1	13.9	15.3	91.2
2000	38.4	38.8	23.9	24.1	22.7	22.9	14.2	14.4	99.0
2005	40.4	40.2	22.6	22.5	22.8	22.7	14.7	14.6	100.5
2006	40.0	40.0	22.2	22.2	22.5	22.5	15.2	15.2	99.9
2007P	39.8	39.2	23.6	23.3	22.8	22.4	15.4	15.1	101.6

Source: EIA, *Annual Energy Review 2007*, Table 1.3.
Notes: Percentages calculated by CRS. Other includes nuclear and renewable energy. Data for 2007 are preliminary.

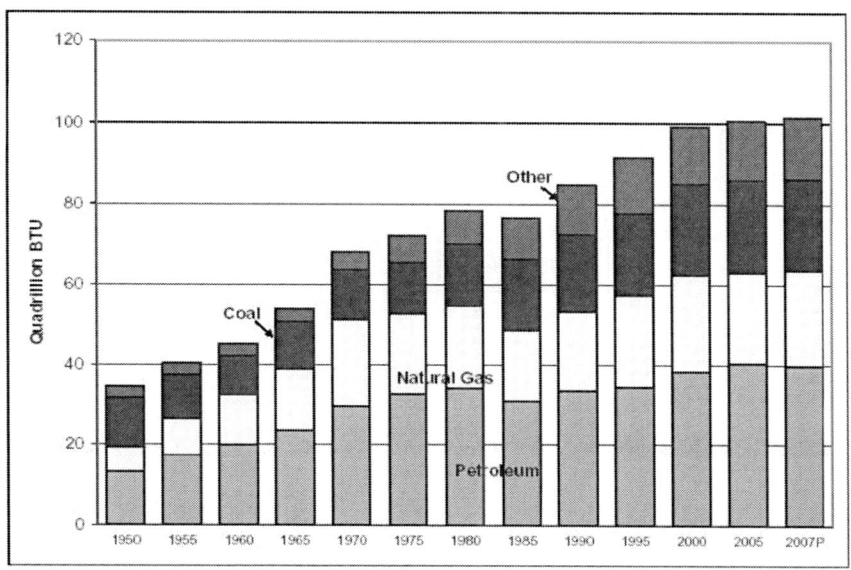

Source: EIA, *Annual Energy Review 2007*, Table 1.3.
Notes: "Other" includes nuclear and renewable energy. Data for 2007 are preliminary.

Figure 3. U.S. Energy Consumption, 1950-2007

OIL

About 40% of the energy consumed in the United States is supplied by petroleum, and that proportion has remained approximately the same since 1950, as the data in the previous section show. Also unchanged is the almost total dependence of the transportation sector on petroleum, mostly gasoline.

The perception that the world is on the verge of running out of oil, widespread during the 1970s, has changed, however. The rapid price increases at that time, aided by improved exploration and production technology, stimulated a global search for oil and resulted in the discovery of large amounts of new reserves. Indeed, as concerns about tightening supply and continually increasing prices were at a peak, proven reserves actually increased by about 50% between 1973 and 1990. Some of the increase was in the Western Hemisphere, mostly in Mexico, but most was located in the region that already dominated the world oil market, the Middle East. With prices essentially steady during the 1990s, the search for oil slowed, but additions to reserves during the decade exceeded the amount of oil pumped out of the ground. By 2003, improved technology for retrieving petroleum from oil sands in Canada and, to a lesser extent, from heavy oil in Venezuela led to significant production from these resources, and by 2005, approximately 200 billion barrels of resources from oil sands and heavy oil were added to the total of proven world reserves, 20% of the total 1991 figure. These trends are illustrated in **Figure 4**.

Petroleum Consumption, Supply, and Imports

Consumption of petroleum by sector reflects a variety of trends (see **Table 3**). In the residential and commercial sectors, petroleum consumption grew steadily from 1950 to 1970, while accounting for about 15% of total petroleum consumption.

Table 3. Petroleum Consumption by Sector, 1950-2007 (Million Barrels per Day (MBD) and Percentage of Total)

	Residential & Commercial		Industrial		Electric		Transportation		Total
	MBD	% of total	MBD	% of total	MBD	% of total	MBD	% of total	MBD
1950	1.1	16.5%	1.8	28.0%	0.2	3.2%	3.4	51.6%	6.5
1955	1.4	16.5%	2.4	28.1%	0.2	2.4%	4.5	52.4%	8.5
1960	1.7	17.5%	2.7	27.6%	0.2	2.5%	5.1	52.4%	9.8
1965	1.9	16.6%	3.2	27.2%	0.3	2.7%	6.0	52.5%	11.5
1970	2.2	14.9%	3.8	25.9%	0.9	6.3%	7.8	52.9%	14.7
1975	1.9	11.9%	4.0	24.8%	1.4	8.5%	9.0	54.9%	16.3
1980	1.5	8.9%	4.8	28.3%	1.2	6.7%	9.5	55.8%	17.1
1985	1.3	8.6%	4.1	25.9%	0.5	3.0%	9.8	62.7%	15.7
1990	1.2	7.2%	4.3	25.3%	0.6	3.3%	10.9	64.0%	17.0
1995	1.1	6.4%	4.6	26.0%	0.3	1.9%	11.7	65.9%	17.7
2000	1.3	6.5%	4.9	24.9%	0.5	2.6%	13.0	66.1%	19.7
2005	1.2	5.8%	5.1	24.5%	0.5	2.6%	14.0	67.1%	20.8
2006	1.1	5.2%	5.1	24.8%	0.3	1.4%	14.2	68.6%	20.7
2007	1.1	5.2%	5.1	24.4%	0.3	1.4%	14.3	68.9%	20.7

Source: EIA, Annual Energy Review 2007, Tables 5.1 and 5.13a-d.
Notes: Percentages calculated by CRS.

After the price surge in the 1970s, consumption in those sectors declined, falling to less than 7% of total petroleum consumption by 1995. When oil prices surged again after 2005, consumption declined further, to about 5%. Usage in the electric power sector followed a similar but more abrupt pattern. Until 1965 only about 3% of petroleum went to power generation. In the late 1960s efforts to improve air quality by reducing emissions led utilities to convert a number of coal-fired power plants to burn oil, and many new plants were designed to burn oil or natural gas. Utilities found themselves committed to increasing dependence on oil just at the time of shortages and high prices; in 1975 almost 9% of oil consumption went for power production. Consumption then fell sharply as alternate sources became available, declining to about 2%-3% of total consumption and falling even lower after 2005 as oil prices increased sharply.

Table 4. U.S. Petroleum Production, 1950-2007 (Million Barrels per Day)

	Crude Oil			Gas Liquids	Other	Total
	48 States	Alaska	Total			
1950	5.4	—	5.4	0.5	—	5.9
1955	6.8	—	6.8	0.8	—	7.6
1960	7	—	7	0.9	0.2	8.1
1965	7.8	—	7.8	1.2	0.2	9.2
1970	9.4	0.2	9.6	1.7	0.4	11.7
1975	8.2	0.2	8.4	1.6	0.5	10.5
1980	7	1.6	8.6	1.6	0.6	10.8
1985	7.2	1.8	9	1.6	0.6	11.1
1990	5.6	1.8	7.4	1.6	0.7	9.6
1995	5.1	1.5	6.6	1.8	0.8	9.1
2000	4.9	1.0	5.8	1.9	1.0	8.7
2005	4.3	0.9	5.2	1.7	1.0	7.9
2006	4.4	0.7	5.1	1.7	1.0	7.8
2007	4.4	0.7	5.1	1.8	1.0	7.9

Source: EIA, Annual Energy Review 2007, Table 5.1.
Notes: "Other" includes processing gain.

While petroleum consumption increased throughout the period from 1950 to the present (except for a temporary decline following the price surge of the 1970s), U.S. domestic production peaked in 1970 (see **Table 4**). The result, as shown in **Figure 5**, was greater dependence on imported petroleum, which rose from less than 20% in 1960 to about 60% in recent years.

Industrial consumption of petroleum, which includes such large consumers as refineries and petrochemical industries, has remained about 25% of total consumption since 1970. As other sectors' share fell, transportation, which was a little more than half of total consumption prior to 1975, climbed to two-thirds by 2000 and continued to increase its share since then.

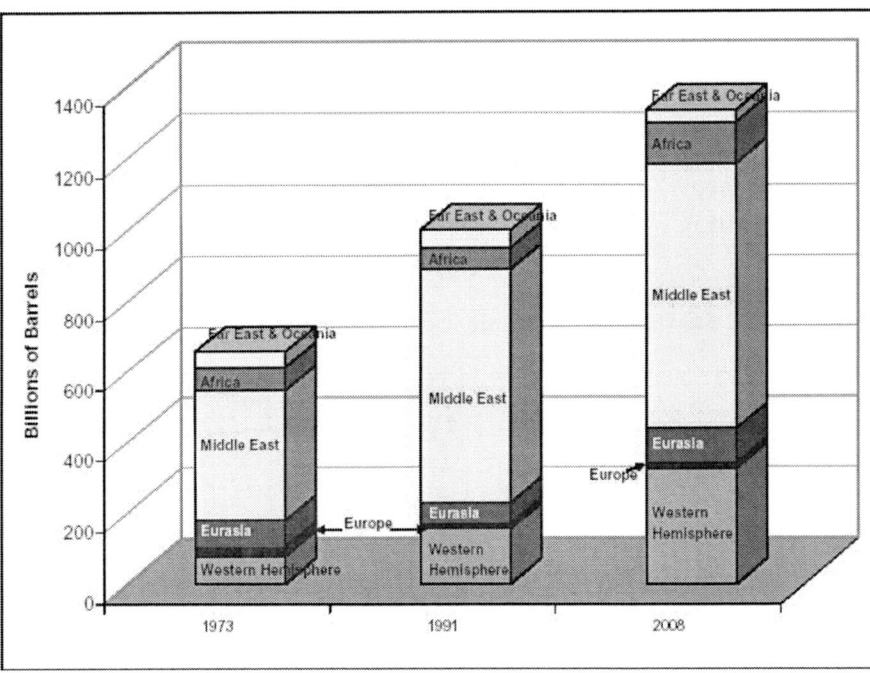

Source: EIA, International Energy Annual (IEA) 1990, Table 32 and IEA 2007 Table 8.1 Table of World Proved Oil and Natural Gas Reserves, Most Recent Estimates. (data is from Oil and Gas Journal and is not certified by EIA, except for the data for the United States in the Western Hemisphere category).

Notes: The categories "Eastern Europe and Former Soviet Union" and "Western Europe," in the data for 1973 and 1991, were changed to "Eurasia" and "Europe" respectively for 2005. Seven countries (Albania, Bulgaria, Czech Republic, Hungary, Poland, Romania, and Slovakia) were moved from the former to the latter.

Figure 4. World Crude Oil Reserves, 1973, 1991 and 2008

U.S. Energy: Overview and Selected Facts and Numbers

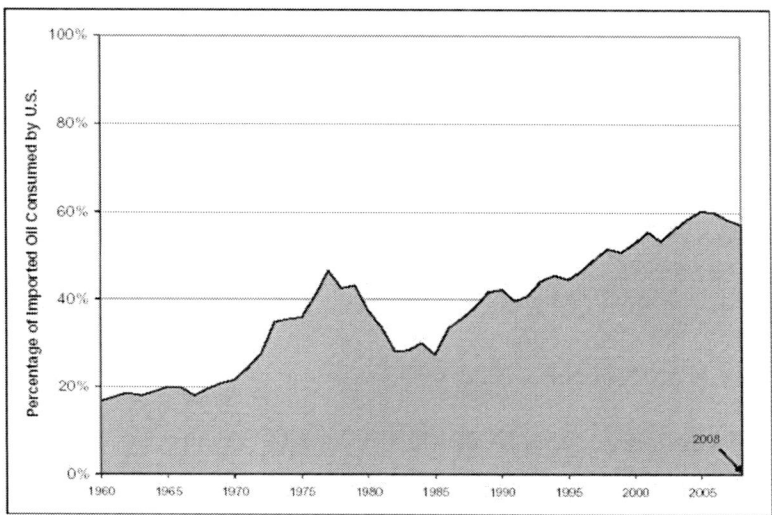

Source: EIA, Monthly Energy Review, December 2008, Table 3.3a, and Annual Energy Review 1986, Table 51.
Notes: Data for 2008 represents an 11-month average.

Figure 5. U.S. Dependence on Imported Petroleum, 1960-2007

**Table 5. Transportation Use of Petroleum, 1950-2007
(Million Barrels Per Day)**

	Aviation	Diesel Fuel	Gasoline	Other	Total
1950	0.1	0.2	2.4	0.6	3.4
1955	0.3	0.4	3.2	0.5	4.5
1960	0.5	0.4	3.7	0.4	5.1
1965	0.7	0.5	4.4	0.4	6.0
1970	1.0	0.7	5.6	0.4	7.8
1975	1.0	1.0	6.5	0.4	9.0
1980	1.1	1.3	6.4	0.7	9.5
1985	1.2	1.5	6.7	0.4	9.8
1990	1.5	1.7	7.1	0.5	10.9
1995	1.5	2.0	7.7	0.5	11.7
2000	1.7	2.4	8.4	0.5	13.0
2005	1.7	2.9	8.9	0.5	14.0
2006	1.7	3.0	9.0	0.5	14.2
2007P	1.6	3.0	9.1	0.5	14.3

Source: EIA, Annual Energy Review 2007, Table 5.13c.
Notes: Data for 2007 are preliminary.

Petroleum and Transportation

Since the transportation sector is so heavily dependent on petroleum, and uses so much of it, **Table 5** and **Figure 6** present a more detailed breakdown of the various types of petroleum used.

Aviation fuel includes both aviation gasoline and kerosene jet fuel. In 1950 aviation was almost entirely gasoline powered; by 2000 it was 99% jet fueled. The growth in flying is illustrated by the fact that aviation fuel was only 3% of petroleum consumption for transportation in 1950, but had grown to 12% in 1965 and has maintained that share since then.

Diesel fuel consumption showed a similar dramatic increase. About 6% of total petroleum consumption for transportation in 1950, it rose to 11% by 1975 and to 20% in recent years. Diesel fuel is used by a number of transportation sectors. Part of the increase involved the change of railroads from coal-fired steam to diesel and diesel-electric power. Diesel fuel is used also in the marine transportation sector, and some private automobiles are diesel-powered. The major part of diesel fuel consumption in transportation is by large commercial trucks. Total diesel fuel consumption increased from about 200,000 barrels per day in 1950 to 3.0 million barrels per day in 2007.

Most of the petroleum consumed in the transportation sector is motor gasoline. In 1950 it was 71% of total sector petroleum consumption, and in recent years, despite the increase in aviation fuel and diesel, it has been about 65%. Since 1950, gasoline consumption has almost quadrupled.

Of the other petroleum products consumed in the transportation sector, the largest is residual fuel oil, most of which is used in large marine transport. Consumption of residual fuel oil in the transportation sector was about 500,000 barrels in 1950, and declined gradually to about 400,000 in 2000.

Petroleum Prices: Historical Trends

Most commodity prices are typically volatile. Because oil is widely consumed, and is so important at all levels of the economy, its price is closely watched and analyzed. Especially since the 1970s, when a generally stable market dominated by a few large oil companies was broken by the Organization of Petroleum Exporting Countries (OPEC) cartel and a relatively open world market came into being, the price of crude oil has been particularly volatile. **Figure 7** and **Figure 8** show the long-term trends of crude oil and gasoline prices, in both current dollars and deflated dollars. The data for these charts do not extend to the collapse in oil prices that began in October 2008. (See "Petroleum Prices: The 2004-2008 Bubble" below.)

At the consumer level, prices of products such as motor gasoline and heating oil have reacted to price and supply disruptions in ways that have been modulated by various government and industry policies and international events. A significant and not often noted fact is that, like many commodities, the long-term trend in gasoline prices, adjusted for inflation and excluding temporary surges, has been down. As shown in **Figure 8**, the real price of gasoline peaked in 1980, then fell precipitously in the mid-1980s. The recent surge in prices brought the price above the peak of 1980 (in real dollars).

Figure 9 illustrates the proportion of the gross domestic product (GDP) dedicated to consumer spending on oil. The price surges in the 1970s pushed this ratio from about 4.5% before the Arab oil embargo to about 8.5% following the crisis in Iran late in the decade. Following that, it declined to less than 4%; during the recent run-up of prices the trend started back up again.

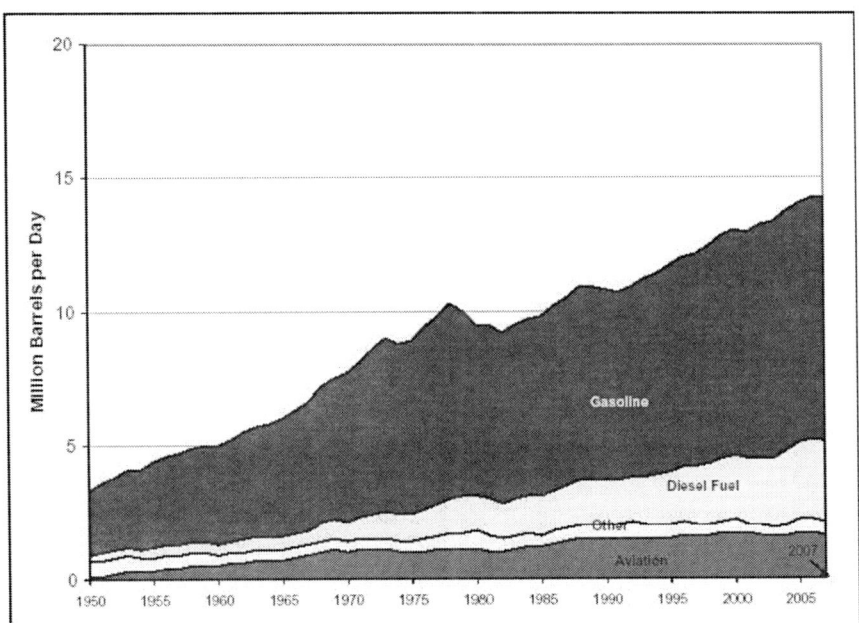

Source: EIA, Annual Energy Review 2007, Table 5.13c.
Notes: Data for 2007 are preliminary.

Figure 6. Transportation Use of Petroleum, 1950-2007

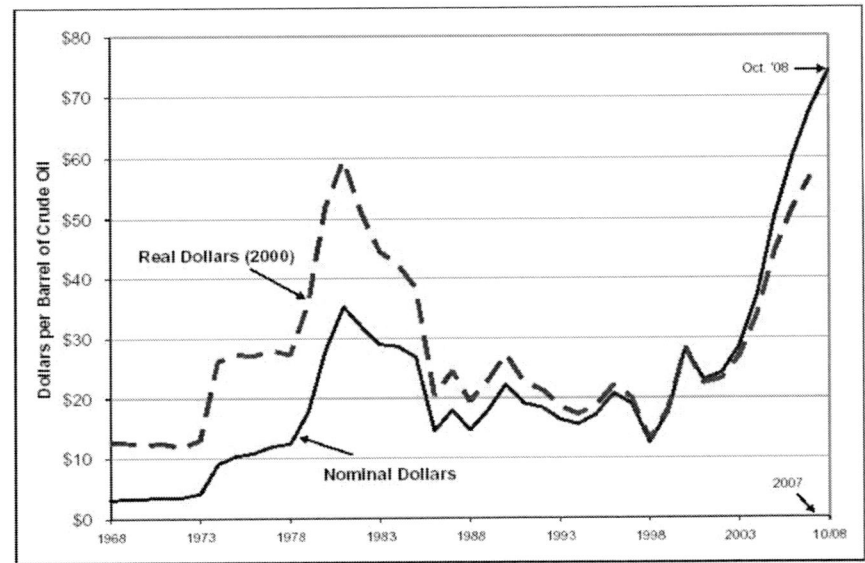

Source: EIA, Annual Energy Review 2007, Table 5.21 and Monthly Energy Review, January 2009, Table 9.1

Notes: Composite crude oil refiner acquisition cost as reported by EIA.

Figure 7. Nominal and Real Cost of Crude Oil to Refiners, 1968-2007 and October 2008

Petroleum Prices: The 2004-2008 Bubble

Beginning in 2004 the world price of crude oil, and with it the price of gasoline, began to increase. Unlike the previous increases in the 1970s, there was no interruption or shortage in the supply of either petroleum or its products, except for a few months in the fall of 2005 when Hurricane Katrina shut down a major portion of U.S. refinery capacity. Nevertheless, an unexpected surge in demand for oil imports to China, added to continuing increases in demand from Europe and the United States as economies continued to grow, tightened the production capacity of the major oil producing nations and signaled that demand in the near future might not be met. In addition, turmoil in the Middle East and elsewhere, as well as the possibility of further natural disasters like Katrina, threatened supply interruptions and put further upward pressure on prices. (See **Figures 10** and **11**.)

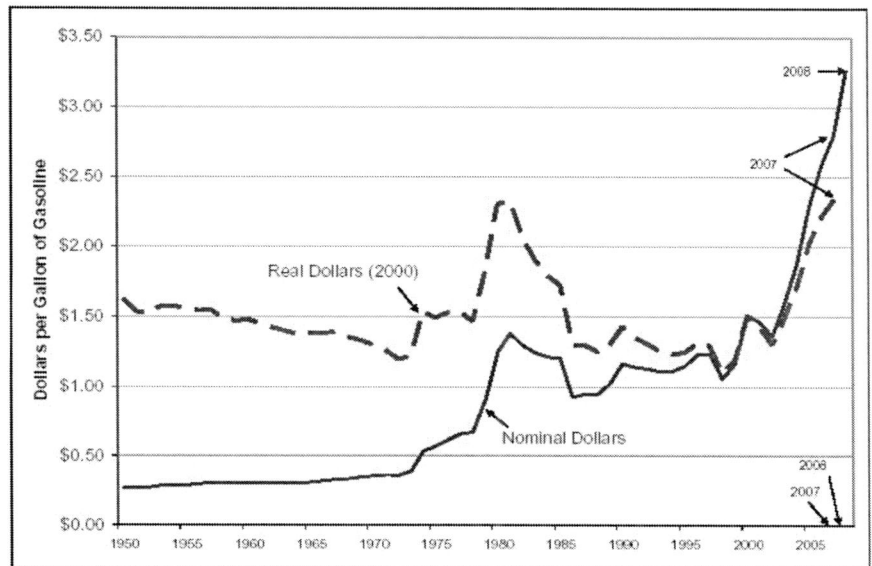

Source: EIA, Annual Energy Review 2007, Table 5.24 and Monthly Energy Review, January 2009, Table 9.4.

Notes: Average national retail price per gallon of unleaded regular gasoline, including taxes.

Figure 8. Nominal and Real Price of Gasoline, 1950-2008

As prices continued to climb, it became apparent that demand for gasoline was relatively insensitive to its cost to the consumer. Throughout the period, as illustrated in **Figure 12**, consumption of gasoline varied seasonally but continued an upward trend on an annual basis. In the summer of 2008 crude oil prices soared far beyond the actual cost of production, and the market took on features of a classical commodities bubble, with expectations of indefinitely rising prices and participation in the market by many who would not normally enter it.

The bubble burst in October 2008 with the onset of a financial crisis in the housing and banking sectors and the evidence that consumption of gasoline was finally faltering. As the economic crisis became more acute, crude prices fell in a few months from $135 per barrel to close to $40, where they had been at the start of the run-up five years earlier.

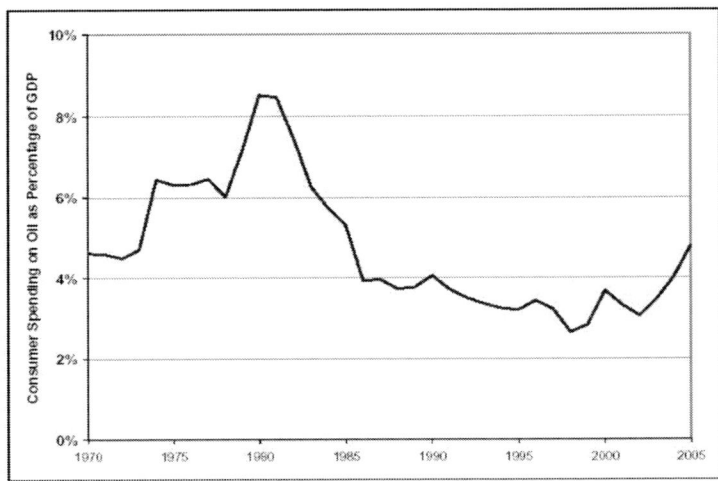

Source: EIA, *Annual Energy Review 2007*, Table 3.5 and Table D1 for GDP in billions of nominal dollars. Percentages calculated by CRS.

Figure 9. Consumer Spending on Oil as a Percentage of GDP, 1970-2005

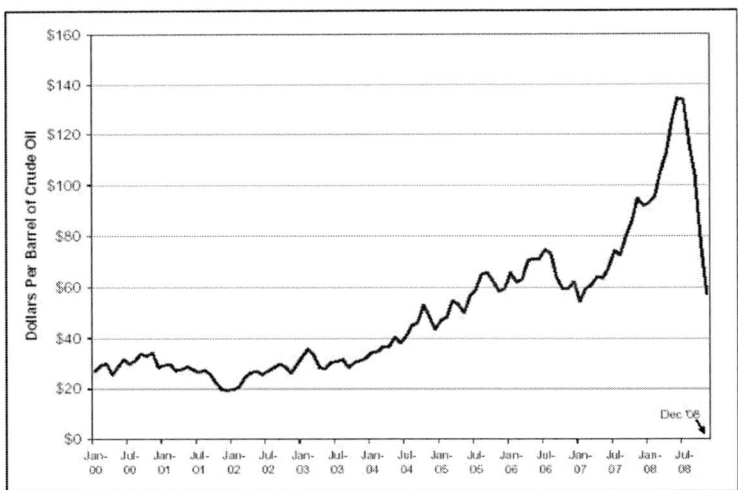

Source: EIA, NYMEX Futures Prices Crude Oil (Light-Sweet, Cushing, Oklahoma) Cushing, OK Crude Oil Future Contract 1

Notes: The futures prices shown are the official daily closing prices at 2:30 p.m. from the trading floor of the New York Mercantile Exchange (NYMEX) for a specific delivery month for each product listed.

Figure 10. Crude Oil Futures Prices.

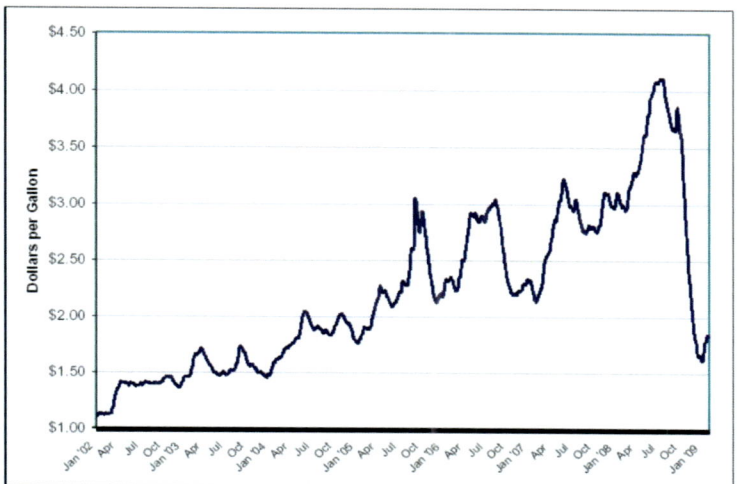

Source: Daily Fuel Gauge Report, American Automobile Association, http://www.fuelgaugereport.com, compiled by CRS.

Notes: Prices include federal, state, and local taxes. Last date above is January 29, 2009; $1.84.

Figure 11. Average Daily Nationwide Price of Unleaded Gasoline, January 2002 – January 2009

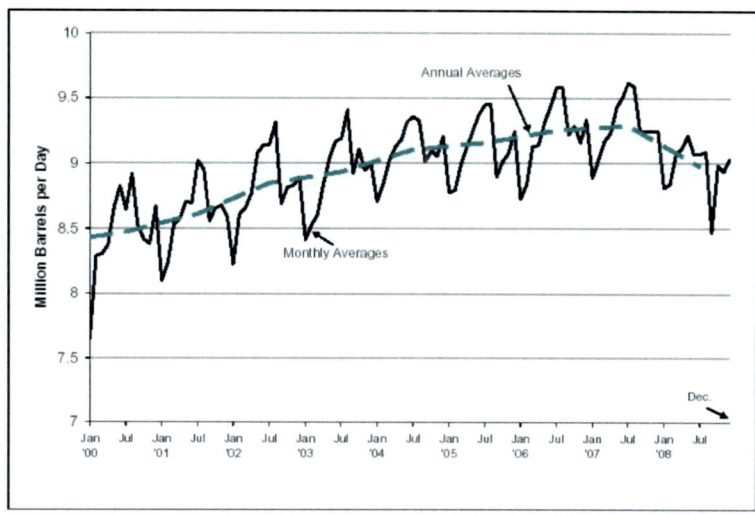

Source: EIA, Monthly Energy Review, January 2009, Table 3.5.

Figure 12. U.S. Gasoline Consumption, January 2000 – December 2008

Gasoline Taxes

The federal tax on gasoline is currently 18.4 cents per gallon. An extensive list of the gasoline and diesel fuel tax rates imposed by each state per gallon of motor fuel is maintained and updated by the American Petroleum Institute (API), "Notes to State Motor Fuel Excise and Other Tax Rates," at [http://www.api.org/policy/tax/stateexcise/upload/December_2007_notes.pdf].

ELECTRICITY

While overall energy consumption in the United States increased nearly three-fold since 1950, electricity consumption increased even more rapidly. Annual power generation is ten times what it was in 1950. **Figure 13** illustrates the trend.

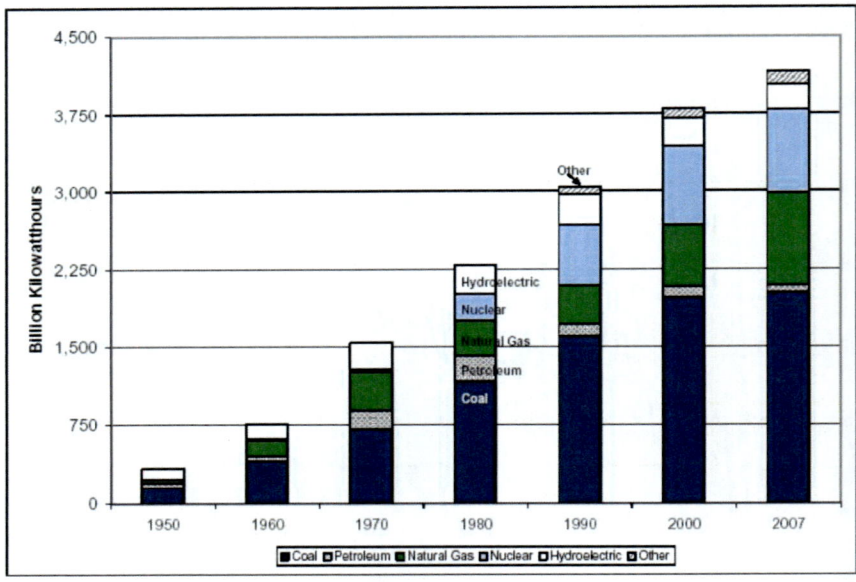

Source: EIA, *Annual Energy Review 2007*, Table 8.2a

Figure 13. Electricity Generation by Source, SelectedYears, 1950-2007

Throughout this period, coal was used to generate about half the rapidly increasing amount of electricity consumed. Petroleum became briefly important as a source of power generation in the late 1960s because it resulted in lower emissions of air pollutants, and consumption continued in the 1970s despite the price surge because natural gas was in short supply. By the 1980s, however, oil consumption by utilities dropped sharply, and in 2007 only 1.2% of power generation was oil-fired.

Natural gas generation has a more complicated history. Consumption by the electric power industry increased gradually as access by pipeline became more widespread. With the price increase in oil in the 1970s, demand for gas also increased, but interstate prices were regulated, and gas availability declined. In addition, federal energy policy viewed generation of electricity by gas to be a wasteful use of a diminishing resource. The Fuel Use Act of 1978 prohibited new power generators from using gas and set a timetable for shutting down existing gas-fired plants. Gas prices were later deregulated, resulting in increased production, and the Fuel Use Act was repealed, but in the meantime generation of electricity from gas fell from 24% in 1970 to 12% in 1985. In the 1990s gas became more popular as technology improved, and by 2000 was supplying 16% of total electric generation. Most capacity additions in the last decade have been gas-fired, as illustrated in **Figure 14**. The increased demand contributed to high prices in 2000 that were felt particularly in California.

Nuclear power started coming on line in significant amounts in the late 1960s, and by 1975, in the midst of the oil crisis, was supplying 9% of total generation. However, increases in capital costs, construction delays, and public opposition to nuclear power following the Three Mile Island accident in 1979 curtailed expansion of the technology, and many construction projects were cancelled. Continuation of some construction increased the nuclear share of generation to 20% in 1990, where it remains currently. Recently, plans have been announced for license applications for up to 30 new reactors, and several have been submitted to the Nuclear Regulatory Commission, but no new plants are currently under construction or on order.

Construction of major hydroelectric projects has also essentially ceased, and hydropower's share of electricity generation has gradually declined from 30% in 1950 to 15% in 1975 and less than 10% in 2000. However, hydropower remains highly important on a regional basis.

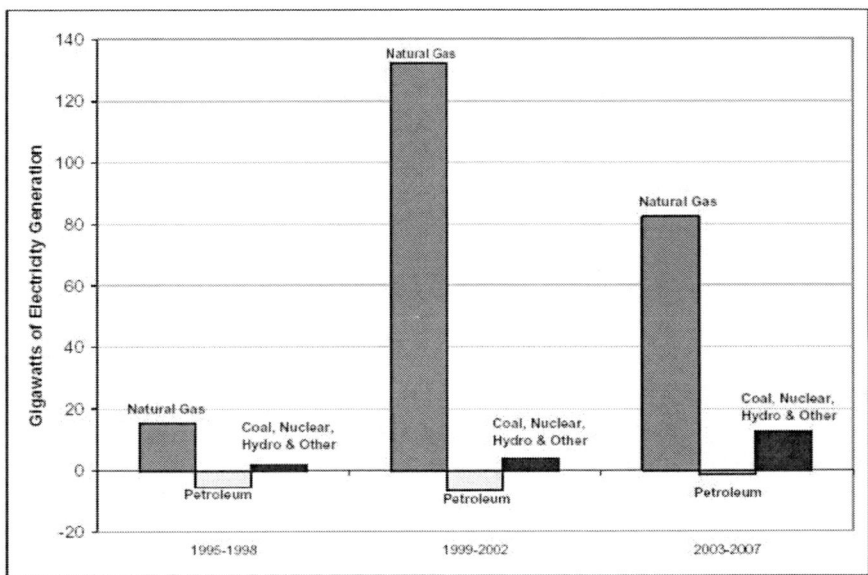

Source: EIA, Annual Energy Review 2007, Table 8.11a.

Figure 14. Changes in Generating Capacity, 1995 - 2007

Sources of power generation vary greatly by region (see **Table 6**). Hydropower in the Pacific Coast states, for instance, supplies over 40% of total generation, and natural gas almost 35%. In 2000, the combination of a drought-caused shortage of hydropower, a tightening of gas supply, and California's new electric regulatory scheme and market manipulation caused very sharp increases in electricity prices in that region. Other regions are heavily dependent on coal generation: The North Central and East South Central states, as well as the Mountain states, generate more than 60% of their electricity from coal, whereas other regions, such as New England and the Pacific Coast, use relatively little coal. The West South Central region (Arkansas, Louisiana, Oklahoma, and Texas) generates 45% of its electricity from gas. New England in the 1970s and 1980s was heavily dependent on oil-generated power; in 2005, despite an increased use of natural gas, oil produced 10% of New England's power, compared with the national average of 2.5%. By 2007, the proportion had dropped to 4.4%, and the national average to 1.2%.

Table 6. Electricity Generation by Region and Fuel, 2007

	Total Generation (billion kwh)	Percentage by					
		Coal	Petroleum	Natural Gas	Nuclear	Hydro	Other
New England	133.5	15.0%	4.4%	40.5%	27.7%	5.7%	6.8%
Middle Atlantic	436.9	35.6%	2.2%	19.2%	34.8%	6.4%	1.8%
East North Central	670.8	69.1%	0.2%	5.3%	23.2%	0.6%	1.6%
West North Central	314.0	74.2%	0.2%	4.8%	15.4%	2.3%	3.1%
South Atlantic	836.5	53.0%	2.4%	17.0%	23.5%	1.4%	2.7%
East South Central	386.9	63.8%	0.2%	12.2%	18.7%	2.9%	2.2%
West South Central	625.7	36.8%	0.1%	45.5%	11.8%	1.2%	4.6%
Mountain	362.9	58.1%	0.1%	24.3%	7.4%	8.2%	1.9%
Pacific Contiguous	373.8	4.1%	0.1%	37.1%	11.7%	37.5%	9.5%
Pacific Noncontiguous	18.6	11.8%	55.4%	21.0%	0.0%	7.1%	4.8%
U.S. Total	4,159.5	48.6%	1.2%	21.5%	19.4%	6.0%	3.4%

Source: EIA, Electric Power Monthly, March 2008, Tables 1.6B, 1.7B, 1.8B, 1.10B, 1.12B, and 1.13B.

Note: "Other" includes renewables other than hydro, plus hydro from pumped storage, petroleum coke, gases other than natural gas, and other sources.

The price of electricity varies by region, depending on the fuel mix and the local regulatory system, among other factors. The nationwide average retail price to residential consumers increased during the 1970s energy crises but declined starting in the 1980s, as indicated by **Figure 15**. An increase starting in 2000 resulted from the expiration in numerous regions of price caps that had been previously imposed when utilities were deregulated; the recent runup in oil and natural gas prices, and to a lesser extent in coal prices, has maintained the trend.

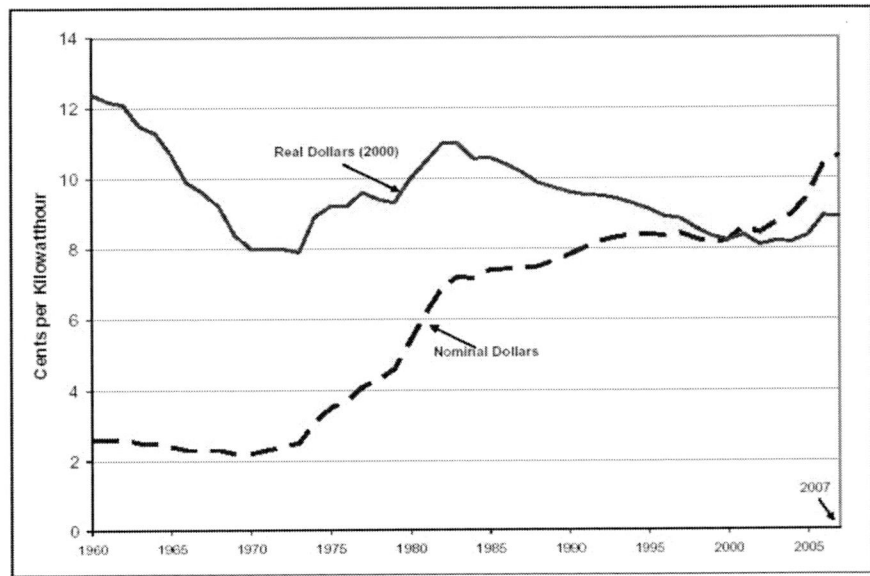

Source: EIA, Annual Energy Review 2007, Table 8.10.
Notes: Price includes taxes.

Figure 15. Price of Retail Residential Electricity, 1960-2007

OTHER CONVENTIONAL ENERGY RESOURCES

Natural Gas

Consumption of natural gas was almost four times as great in 2007 as it was in 1950. Throughout the period, consumption in the residential and commercial sector grew at about the same rate as total consumption, in the range of 30% to 40% of the total. As shown in **Table 7**, consumption for electric power generation increased from about 10% in 1950 to more than 20% at the end of the century. The proportion of total gas consumption by the industrial sector declined correspondingly, from more than 50% in 1950 to about 35% in recent years.

In part because of increased demand by electric utilities, natural gas prices have become extremely volatile in recent years, as illustrated by **Figure 16**, which shows high, low, and yearly average prices for gas delivered to electricity generators.

Table 7. Natural Gas Consumption by Sector, 1950-2007

	Total Consumption trillion cubic feet (tcf)	Percent Consumed by Sector		
		Residential - Commercial	Industrial	Electric
1950	5.77	27.5%	59.4%	10.9%
1955	8.69	31.7%	52.2%	13.3%
1960	11.97	34.5%	48.2%	14.4%
1965	15.28	35.0%	46.5%	15.2%
1970	21.14	34.2%	43.8%	18.6%
1975	19.54	38.0%	42.8%	16.2%
1980	19.88	37.0%	41.2%	18.5%
1985	17.28	39.7%	39.7%	17.6%
1990	19.17	36.6%	43.1%	16.9%
1995	22.21	35.5%	42.3%	19.1%
2000	23.33	35.0%	39.8%	22.3%
2001	22.24	35.0%	38.1%	24.0%
2002	23.01	34.9%	37.5%	24.6%
2003	22.38	37.1%	36.9%	22.9%
2004	22.39	35.7%	37.3%	24.4%
2005	22.01	35.6%	35.0%	26.7%
2006	21.65	33.3%	35.2%	28.7%
2007P	23.06	33.5%	33.8%	29.8%

Source: EIA, Annual Energy Review 2007, Table 6.5.
Notes: Data for 2007 are preliminary. Percentages do not add to 100. The remaining amount is used by the transportation sector.

Because rates for residential natural gas are regulated, they have been less volatile than those for electric utility consumers, although considerable seasonal fluctuations are common, as shown in **Figure 17**. The long-term trend in residential natural gas prices, both in current dollars and in constant 2006 dollars, is shown in **Figure 18**.

COAL

Consumption of coal has more than doubled since 1950, but during that period coal as an energy source changed from a widely used resource to a single-use fuel for generating electricity. (See **Table 7**.) In 1950 the residential and commercial sector consumed almost a quarter of total coal consumed; by 1980 less than 1% of coal went to that sector. In transportation, steam locomotives (and some coal-fired marine transportation) consumed 13% of coal; by 1970 they were all replaced with diesel-burning or electric engines. Industry consumed 46% of coal in 1950; by 2000 less than 10% of coal was consumed by that sector. Meanwhile, the electric power sector, which consumed less than 20% of the half-billion tons of coal burned in 1950, used more than 90% of the billion-plus tons consumed in 2007.

RENEWABLES

The major supply of renewable energy in the United States, not counting hydroelectric power generation, is fuel ethanol. Consumption in the United States in 2007 was 6.5 billion gallons, mainly blended into E10 gasohol (a blend of 10% ethanol and 90% gasoline). This figure represents 4.5% of the approximately 140 billion gallons of gasoline consumption in the same year. As **Figure 19** indicates, fuel ethanol production has increased rapidly in recent years, especially since the phasing out of the fuel additive methyl tertiary butyl ether (MTBE).

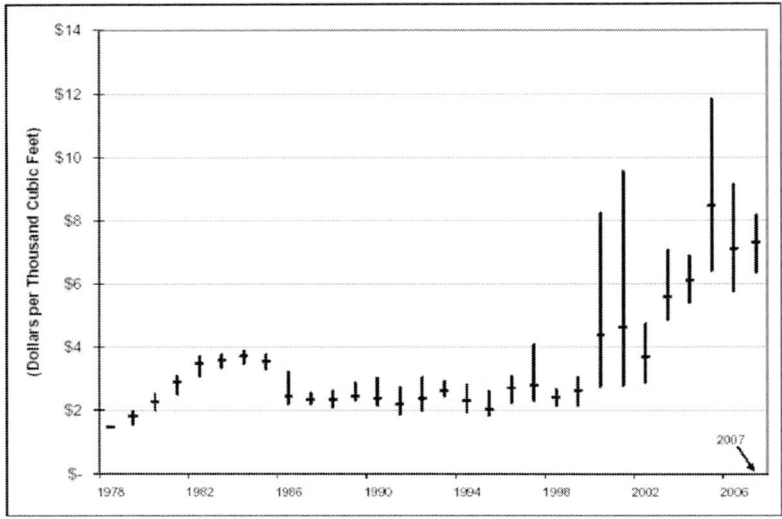

Source: EIA, Monthly Energy Review, December 2008, Table 9.11.

Figure 16. Natural Gas Prices to Electricity Generators, 1978 - 2007

U.S. Energy: Overview and Selected Facts and Numbers 49

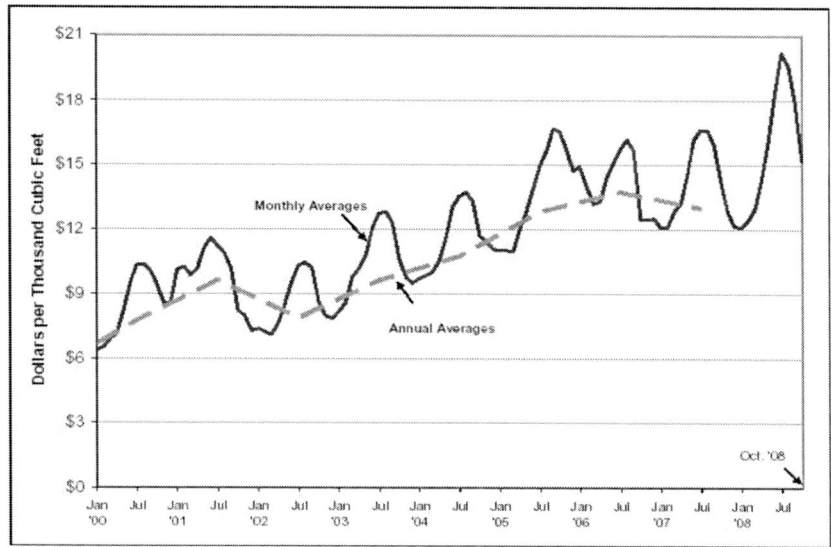

Source: EIA, Monthly Energy Review, January 2009, Table 9.11.

Figure 17. Monthly and Annual Residential Natural Gas Prices, 2000 – October 2008

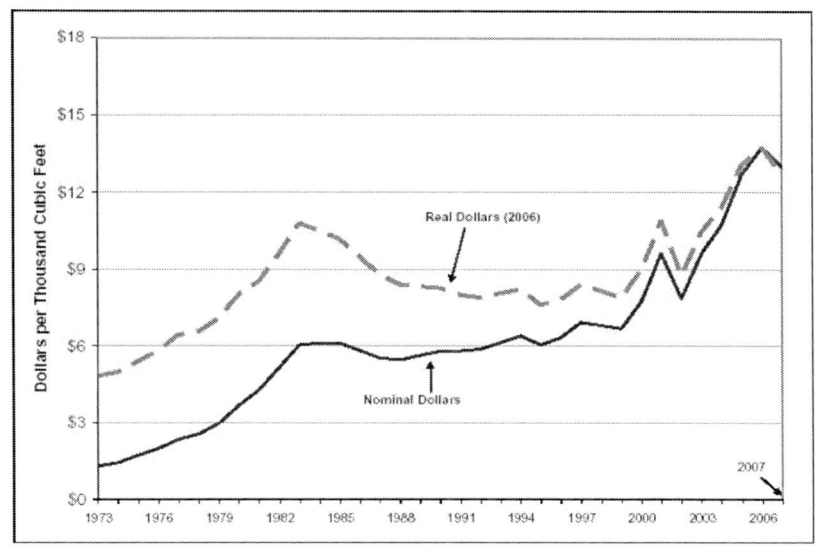

Source: EIA, Monthly Energy Review, December 2008, Table 9.11 and FY2008 Budget, Historical Tables, Table 10.1 for GDP Chained Price Index.

Figure 18. Annual Residential Natural Gas Prices, 1973-2007

Table 8. Coal Consumption by Sector, 1950-2007

	Total Consumption (million tons)	Percent Consumed by Sector			
		Residential-Commercial	Industrial	Transportation	Electric
1950	494.1	23.2%	45.5%	12.8%	18.6%
1955	447.0	15.3%	48.7%	3.8%	32.2%
1960	398.1	10.3%	44.6%	0.8%	44.4%
1965	472.0	5.4%	42.6%	0.1%	51.9%
1970	523.2	3.1%	35.7%	0.1%	61.2%
1975	562.6	1.7%	26.2%	–	72.2%
1980	702.7	0.9%	18.1%	–	81.0%
1985	818.0	1.0%	14.2%	–	84.8%
1990	904.5	0.7%	12.7%	–	86.5%
1995	962.1	0.6%	11.0%	–	88.4%
2000	1084.1	0.4%	8.7%	–	90.9%
2005	1126.0	0.4%	7.4%	–	92.1%
2006	1112.3	0.3%	7.4%	–	92.3%
2007	1128.8	0.3%	7.0%	–	92.7%

Source: EIA, Annual Energy Review 2007, Table 7.3
Notes: Data for 2007 are preliminary.

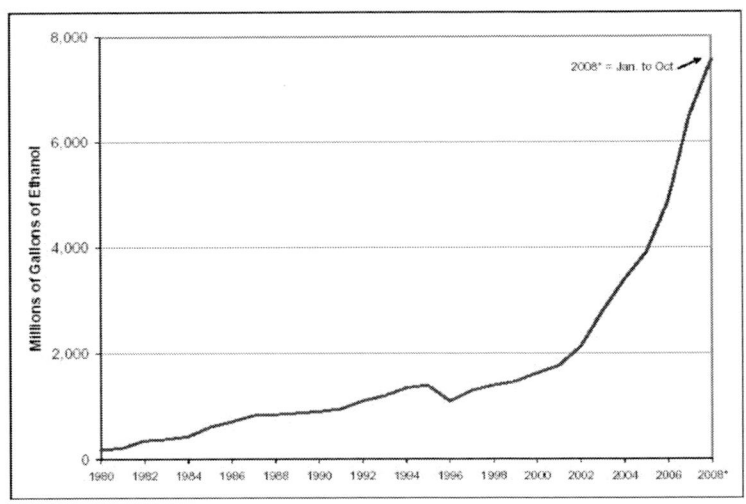

Source: Renewable Fuels Association, January 26, 2008. [http://www.ethanolrfa.org/industry/statistics/].
Notes: Data point for 2008 is for the first ten months of the year.

Figure 19. U.S. Ethanol Production, 1980-2007 and Jan.-Oct. 2008

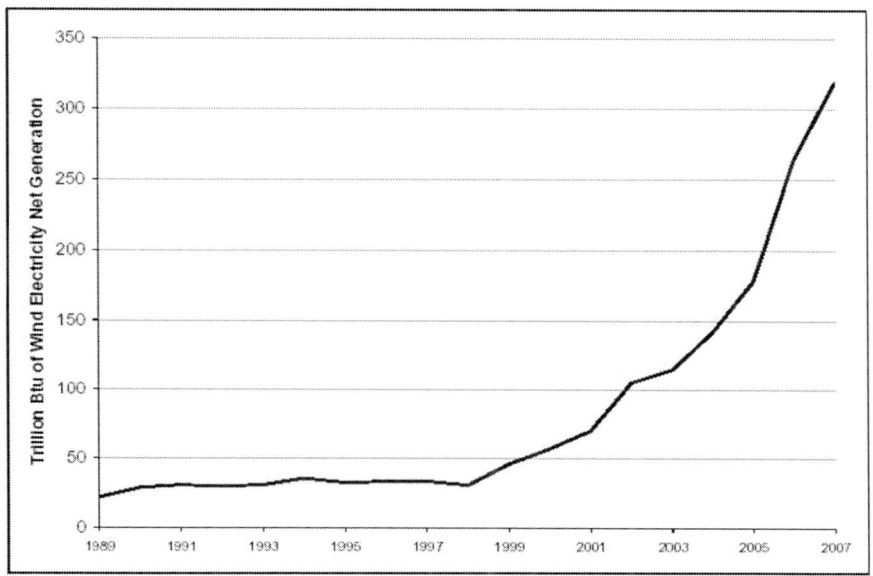

Source: Monthly Energy Review, December 2008, Table 10.1.
Notes: Wind electricity net generation converted to Btu using the fossil-fueled plants heat rate.

Figure 20. Wind Electricity Net Generation, 1989-2007

Another rapidly growing renewable resource is wind-generated electric power, as shown in **Figure 20**. The 300 trillion Btus of wind energy in 2006 is equivalent to approximately 88 billion kilowatt hours, about 2% of the 4,000 billion kwh of total electricity generation in that year.

CONSERVATION AND ENERGY EFFICIENCY

Vehicle Fuel Economy

Energy efficiency has been a popular goal of policy makers in responding to the repeated energy crises of recent decades, and efforts to reduce the energy intensity of a broad spectrum of economic activities have been made both at the government and private level. Because of the transportation sector's near total dependence on vulnerable oil supplies, improving the efficiency of motor vehicles has been of particular interest. (For an analysis of legislative policies to improve vehicle fuel economy, see CRS Report R40166, *Automobile and Light Truck Fuel Economy: The CAFE Standards*, by Brent D. Yacobucci and Robert Bamberger.) **Figure 21** illustrates the trends in this effort for passenger cars and for light trucks, vans, and sport utility vehicles, as well as the general lack of improvement in heavy trucks.

Further analysis by the Environmental Protection Agency (EPA), involving the composition of the fleet as well as the per-vehicle fuel rates, indicates that light vehicle fuel economy has declined on average between 1988 and 2003. This is largely because of increased weight, higher performance, and a higher proportion of sport utility vehicles and light trucks sold. In 2003, SUVs, pickups, and vans comprised 48% of all sales, more than twice their market share in 1983. (The EPA study is available online at [http://www.epa.gov/otaq/fetrends.htm].)

Energy Consumption and GDP

A frequent point of concern in formulating energy policy is the relationship between economic growth and energy use. It seems obvious that greater economic activity would bring with it increased energy consumption, although many other factors affecting consumption make the short-term relationship highly variable. Over a longer period, for some energy-related activities, the relationship with economic growth has been essentially level. For the period from 1973 to 2003, for instance, consumption of electricity remained close to 0.45 kwh per constant dollar of GDP. Similarly, the number of miles driven by all vehicles was close to 3 miles per constant dollar of GDP throughout the same period.

In the case of oil and gas, however, a remarkable drop took place in the ratio of consumption to economic growth following the price spikes and supply disruptions, as illustrated in **Figure 22**. Consumption of oil and gas declined from 14,000 Btus per constant dollar of GDP in 1973 to a little more than 8,000 in 1985, and has continued to decline at a slower rate since then.

During the earlier period, oil and gas consumption actually declined 15% while GDP, despite many economic problems with inflation and slow growth, was increasing by 44% (see **Figure 22**). During the period 1987 to 2007, oil and gas consumption increased by about 25%, while GDP increased 76%.

U.S. Energy: Overview and Selected Facts and Numbers 53

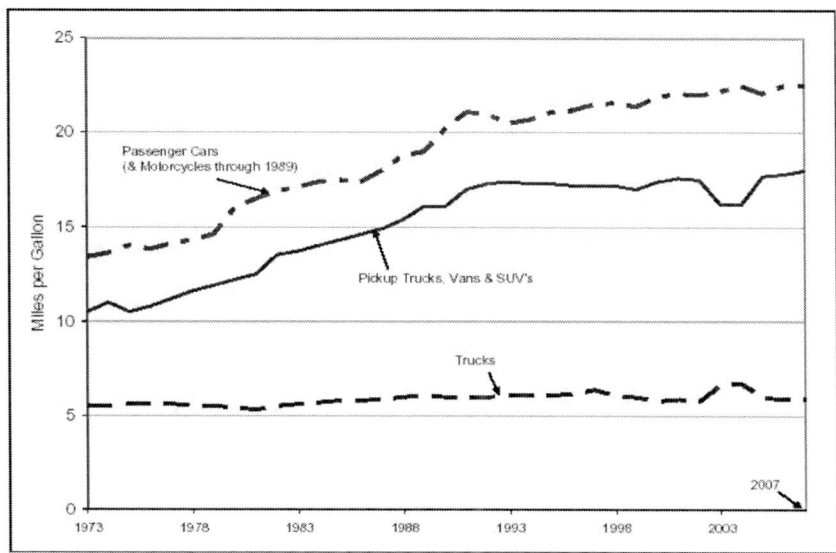

Source: EIA, Monthly Energy Review, December 2008, Table 1.8.

Figure 21. Motor Vehicle Efficiency Rates, 1973-2007

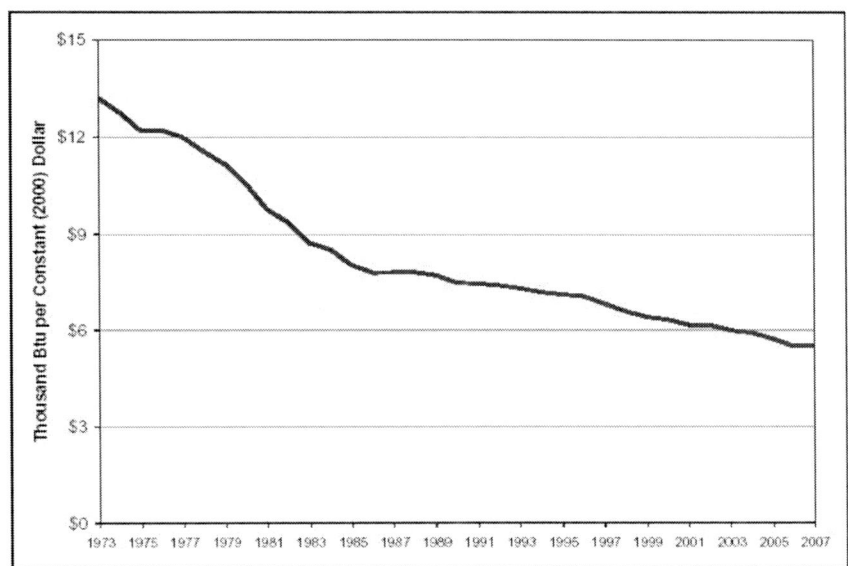

Source: EIA, Monthly Energy Review, January 2009, Table 1.7.

Figure 22. Oil and Natural Gas Consumption per Dollar of GDP, 1973 - 2007

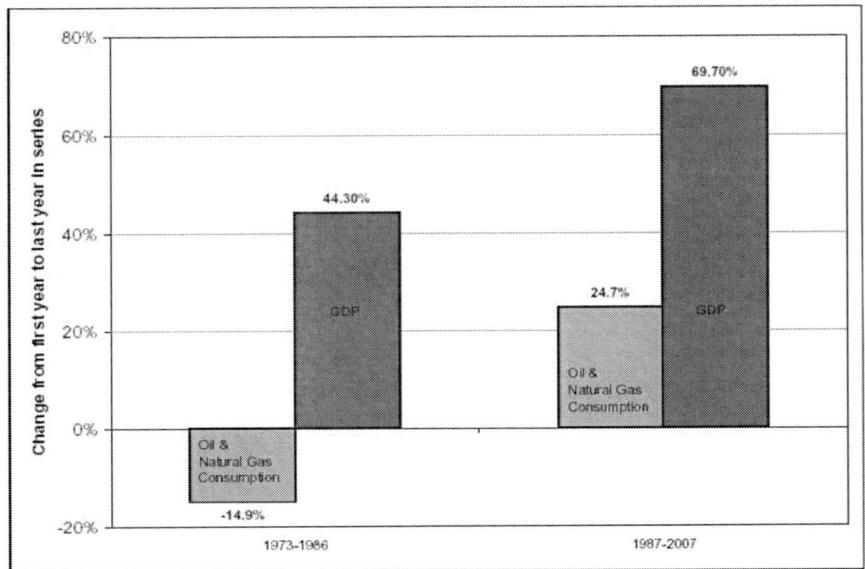

Source: EIA, Monthly Energy Review, January 2009, Table 1.8.
Notes: Percentages calculated by CRS. Percent change in oil & natural gas consumption measured in quadrillion Btu. Percent change in GDP based on billion chained (2000) dollars.

Figure 23. Change in Oil and Natural Gas Consumption and Growth in GDP, 1973-2007

MAJOR STATISTICAL RESOURCES

Energy Information Administration (EIA)

EIA home page—http://www.eia.doe.gov
Most of the tables and figures in this report are derived from databases maintained by the Energy Information Administration (EIA), an independent agency of the Department of Energy. EIA's Website presents the complete text of its many statistical reports in PDF's and Excel files.

EIA, Publications and Reports—http://www.eia.doe.gov/bookshelf.html
EIA's most frequently requested reports include the following:

Annual Energy Review: all the historical yearly energy data across fuels
Annual Energy Outlook: energy projections out to 2030
Country Analysis Briefs: country-level energy overviews
Electric Power Monthly: monthly summary of electric power generation and capacity
International Energy Annual: international historical yearly energy data across fuels
International Energy Outlook: worldwide energy projections to 2025
Monthly Energy Review: all the latest monthly energy data across fuels
This Week in Petroleum: weekly prices & analytical summary of the petroleum industry
Weekly Petroleum Status Report: weekly petroleum prices, production & stocks data

Other Sources

Nuclear Regulatory Commission Information Digest: *http://www.nrc.gov/reading-rm/doc-collections/nuregs/staff/sr1350/*
Updated annually, this official NRC publication (NUREG-1350) includes general statistics on U.S. and worldwide nuclear power production, U.S. nuclear reactors, and radioactive waste

American Petroleum Institute (API): http://api-ec.api.org/newsplashpage/index.cfm
The primary trade association of the oil and natural gas industry representing more than 400 members. Research, programs, and publications on public policy, technical standards, industry statistics, and regulations.

API: State Gasoline Tax Reports: http://www.api.org/statistics

Bloomberg.Com, Market Data: Commodities, Energy Prices: *http://www.bloomberg.com/energy/index.html*
Displays four tables:

- Petroleum ($/bbl) for crude oil. The generally accepted price for crude oil is "WTI Cushing $" which is listed third in the table.
- Petroleum (¢/gal) for heating oil and gasoline.
- Natural Gas ($/MMBtu)
- Electricity ($/megawatt hour)

This site is updated two to three times per day.

AAA's Daily Fuel Gauge Report: http://www.fuelgaugereport.com/index.asp

At-the-pump retail fuel prices for gasoline and diesel fuel. Gives average price for today, yesterday, a month ago and a year ago for wholesale and crude oil. Also displays line chart showing the averages for the previous 12 months. National, state, and metropolitan data.

International Energy Agency: http://www.iea.org

The International Energy Agency is an autonomous body within the Organization for Economic Co-operation and Development (OECD). It gathers and analyzes statistics and "disseminates information on the world energy market and seeks to promote stable international trade in energy."

A subscription is required to access most of the information on this Website, although a limited amount of information is available to nonsubscribers. Members of Congress and their staff should contact CRS for a copy of anything that requires a subscription.

End Notes

[1] Further complications can result from the fact that not all sources use the same abbreviations for the various units. The Energy Information Administration (EIA), for example, abbreviates "million barrels per day" as "MMbbl/d" rather than "mbd." For a list of EIA's abbreviation forms for energy terms, see http://www.eia.doe.gov/neic/a-z/a-z_abbrev/a-z_abbrev.html.

[2] In calculating these percentages, "electric energy consumption" includes both the energy value of the kilowatt-hours consumed and the energy lost in generating that electricity.

CHAPTER SOURCES

The following chapters have been previously published:

Chapter 1 – These remarks were delivered as Written Testimony of Dr. Fatih Birol, Chief Economist, International Energy Agency, before the Subcommittee on Energy and Mineral Resources, Committee on Natural Resources, U.S. House of Representatives, dated March 5, 2009.

Chapter 2 – These remarks were delivered as Statement of Dr. Howard Gruenspecht, Acting Administrator, Energy Information Administration, before the Subcommittee on Energy and Mineral Resources, Committee on Natural Resources, U.S. House of Representatives, dated March 5, 2009.

Chapter 3 – These remarks were delivered as Statement of Brenda S. Pierce, Program Coordinator, Energy Resources Program, U.S. Geological Survey, U.S. Department of the Interior, before the Subcommittee on Energy and Mineral Resources, Committee on Natural Resources, U.S. House of Representatives, dated March 5, 2009.

Chapter 4 - This is an edited, excerpted and augmented edition of a United States Congressional Research Service publication, Report Order Code R40187, dated February 3, 2009.

INDEX

A

aid, 25
air, 32, 43
air quality, 32
Alaska, 6, 9, 16, 17, 34
alternative, 7, 12, 19
API, 42, 55
application, 20
Arctic, 23
Arkansas, 44
assessment, 2, 16, 17, 18, 19, 20, 21, 23, 24
assumptions, 2, 7, 8, 13
Atlantic, 9, 10, 11, 13, 45
automobiles, 36
availability, 6, 12, 16, 43
aviation, 36

B

back, 3, 37
banking, 39
biofuels, 6
biomass, 7
breakdown, 36
broad spectrum, 51
Btus, 27, 51, 52
bubble, 39
Bureau of Land Management, 17
burning, 48
butyl ether, 48

C

capital cost, viii, 26, 43
caps, 45
carbon, 2, 3, 4, 6
cartel, 36
clean energy, 1, 3
climate change, 1, 4, 23
CO2, 4
coal, viii, 6, 16, 21, 24, 26, 27, 30, 32, 36, 43, 44, 45, 47
collaboration, 1, 4
Colorado, 20
combined effect, 6
commodities, 37, 39
commodity, 36
complications, 57
composition, 52
concentration, 2, 19
Congress, 19, 24, 25, 56
constraints, 6, 12
construction, viii, 26, 43
consumers, 1, 32, 45, 47
consumption, vii, viii, 2, 6, 7, 8, 15, 25, 26, 27, 32, 36, 39, 42, 43, 46, 48, 52, 54, 57
conversion, 9
Corporate Average Fuel Economy, 6
costs, viii, 10, 21, 26, 43
credit, 3
CRS, 28, 29, 30, 34, 40, 41, 52, 54, 56

crude oil, 2, 6, 8, 10, 12, 13, 36, 38, 39, 55, 56
crust, 19

D

danger, 3
decision makers, 16, 19, 24
decisions, 6, 16
delivery, 40
density, 23
Department of Energy, 5, 17, 54
Department of the Interior, 8, 16, 59
deposits, 8, 20
diesel, 3, 7, 36, 42, 48, 56
discourse, 24
distribution, 12, 16, 23
domestic crude, 13
drought, 44

E

Eastern Europe, 33
economic activity, 52
economic crisis, 3, 39
economic problem, 52
ecosystem, 24
electric energy, 57
electric power, viii, 6, 7, 22, 26, 32, 36, 43, 46, 48, 51, 55
electricity, viii, 7, 21, 22, 26, 27, 42, 43, 44, 45, 46, 47, 51, 52, 57
embargo, 37
energy, vii, viii, 1, 2, 3, 4, 5, 6, 8, 15, 16, 17, 18, 19, 21, 22, 23, 24, 25, 26, 27, 30, 31, 42, 43, 45, 47, 51, 52, 55, 56, 57
energy consumption, viii, 15, 26, 27, 28, 42, 52, 57
Energy Independence and Security Act, 7
Energy Information Administration (EIA), 5, 7, 28, 29, 54, 57, 59
energy markets, vii, 1, 25, 26
Energy Policy Act of 2005, 11, 18, 19
Energy Policy and Conservation Act, 17
energy supply, 2, 17
engines, 48
environment, 22

environmental conditions, 23
Environmental Protection Agency, 52
EPA, 52
ethanol, viii, 7, 26, 27, 48
Ethanol, 50
exaggeration, 1

F

feet, 7, 8, 9, 10, 11, 13, 16, 18, 20, 23, 27, 47
financial crisis, 39
fire, 6, 32, 36
floating, 12
flow, 10
fluctuations, 47
focusing, 5
Forest Service, 17
fossil, 2, 3, 4, 21, 51
fossil fuel, 3, 4
fuel, viii, 3, 6, 7, 21, 26, 27, 36, 42, 45, 47, 48, 52, 56, 57
funding, 20
futures, 40

G

gas, viii, 2, 3, 6, 7, 8, 9, 10, 11, 12, 13, 16, 17, 18, 19, 20, 22, 23, 24, 25, 26, 27, 28, 30, 32, 43, 44, 45, 46, 47, 52, 54, 55
gases, 45
gasoline, vii, 3, 7, 25, 27, 31, 36, 37, 38, 39, 42, 48, 55, 56
GDP, 37, 40, 49, 52, 53, 54
generation, viii, 6, 7, 21, 22, 26, 27, 32, 42, 43, 44, 46, 48, 51, 55
generators, 7, 43, 46
geology, 16
geophysical, 19
geothermal, 7, 16, 21, 24
global demand, 6
global economy, 6
government, 2, 16, 20, 23,24, 37, 51
greenhouse, 2, 3, 6
greenhouse gas, 2, 3, 6
gross domestic product, 37
groups, 16, 23

growth, 2, 3, 6, 7, 16, 19, 20, 27, 36, 52
Gulf of Mexico, 8, 18

H

health, 15, 16, 24
heat, 51
heating, 27, 37, 55
heavy oil, 32
historical trends, vii, 10, 25, 30
House, vii, 59
housing, 39
human, 2, 15, 16, 24
Hurricane Katrina, 38
hydrates, 16, 18, 20, 24
hydro, 19, 45
hydrocarbon, 19, 24
hydroelectric power, 48
hydropower, viii, 26, 27, 43, 44

I

IEA, 1, 2, 3, 4, 23, 33
imports, viii, 2, 4, 23, 26, 27, 38
increased access, 9
indicators, 11, 12, 26
industrial, 27, 28, 46
industry, vii, 2, 16, 17, 18, 19, 20, 23, 25, 27, 28, 37, 43, 50, 55
inflation, 37, 52
infrastructure, 10
International Energy Agency, 1, 23, 56, 59
international trade, 56
investment, 3, 6

J

jet fuel, 36
jobs, 3
jurisdiction, 18

K

Katrina, 38
kerosene, 36

L

land, 16, 17, 24

laws, 9
light trucks, 52
liquids, 6, 7, 23
Louisiana, 44

M

management, 17
manipulation, 44
market, vii, 1, 4, 18, 19, 21, 25, 31, 36, 39, 44, 52, 56
measures, 1
megawatt, 55
methane, 21
methyl tertiary, 48
metric, 6
Mexico, 2, 18, 31
million barrels per day, 6, 8, 10, 12, 13, 27, 36, 57
mineral resources, 18, 19
minerals, 16, 18
Minerals Management Service (MMS), 7, 8, 15
mining, 21
missions, 3, 4, 6
MMS, 8, 9, 11, 12, 15, 16, 17, 19, 20, 22, 24
models, 21
movement, 26
MTBE, 48

N

nation, 3
national security, 23
natural, viii, 3, 6, 7, 8, 9, 10, 11, 12, 13, 16, 17, 18, 19, 20, 22, 23, 26, 27, 30, 32, 38, 43, 44, 45, 46, 47, 54, 55
natural gas, viii, 3, 6, 7, 8, 9, 10, 11, 12, 13, 16, 17, 18, 19, 20, 22, 23, 26, 27, 30, 32, 43, 44, 45, 46, 47, 54, 55
natural resources, 16
New England, 44, 45
New York, 40
New York Mercantile Exchange, 40
nongovernmental, 16
NRC, 55
nuclear, viii, 4, 26, 27, 30, 31, 43, 55

Nuclear Regulatory Commission, 43, 55
NYMEX, 40

O

oceans, 22
OECD, 2, 4, 56
offshore, 7, 8, 9, 10, 11, 12, 15, 16, 17, 18, 22, 23
offshore oil, 8, 9
oil, vii, viii, 1, 2, 3, 4, 6, 7, 8, 9, 10, 11, 12, 13, 16, 17, 18, 19, 20, 22, 23, 24, 25, 26, 27, 28, 30, 31, 32, 36, 37, 38, 39, 43, 44, 45, 51, 52, 54, 55, 56
Oil and Gas Journal, 33
oil production, 6, 8, 10, 18, 19
oil recovery, 19
oil sands, 32
oil shale, 16, 20, 24
Oklahoma, 40, 44
online, viii, 21, 26, 52
OPEC, 1, 2, 36
opposition, viii, 26, 43
Organization for Economic Cooperation and Development, 6

P

Pacific, 8, 9, 10, 11, 13, 44, 45
passenger, 52
PCA, 17
per capita, 27
perception, 31
perceptions, vii, 25
periodic, 19
petrochemical, 32
Petroleum, 7, 30, 32, 33, 34, 35, 36, 37, 38, 42, 43, 45, 55
petroleum products, 36
pipelines, 12
plants, viii, 6, 26, 27, 32, 43, 51
play, 8, 19, 24
policy makers, 51
policymakers, 5, 24, 25
political stability, 15
population, 22

power, viii, 2, 3, 6, 7, 21, 26, 27, 32, 36, 42, 43, 44, 46, 48, 51, 55
power generation, 22, 32, 42, 43, 44, 48
power plants, 6, 32
pressure, 38
price caps, 45
prices, vii, 3, 6, 8, 10, 13, 20, 21, 25, 31, 32, 36, 37, 38, 39, 40, 43, 44, 45, 46, 47, 55, 56
private, 36, 51
probability, 9, 12
production, viii, 2, 3, 5, 6, 7, 8, 9, 10, 11, 12, 13, 16, 18, 19, 21, 22, 26, 27, 31, 32, 38, 39, 43, 48, 55
production technology, 31
program, 9, 22
prosperity, 2, 15
public, viii, 2, 5, 16, 24, 26, 43, 55
public policy, 55

R

radioactive waste, 55
range, 46
reading, 55
recovery, 3, 19, 20
refineries, 3, 32, 38
regional, viii, 10, 26, 43
regular, 39
regulations, 6, 9, 20, 55
Regulatory Commission, 43, 55
relationship, 11, 52
renewable energy, viii, 22, 26, 30, 31, 48
renewable resource, 21, 51
reserves, 2, 8, 9, 10, 11, 17, 19, 20, 21, 23, 31
reservoirs, 12, 20
residential, 27, 28, 32, 45, 46, 47
resolution, 10
resource management, 17
resources, 3, 6, 7, 8, 9, 12, 13, 15, 16, 17, 18, 19, 20, 21, 22, 23, 24, 28, 30, 32
responsibilities, 19
retail, 39, 45, 56
returns, 11
risk, 3

robustness, 17

S

sales, 21, 52
scaling, 3
scientific understanding, 17
search, 31
security, 3, 4, 23
sensitivity, 7, 12
shares, 18
short supply, 43
shortage, 3, 38, 44
short-term, 52
simulations, 21
solar, 7
spectrum, 51
stability, 15
standards, 3, 6, 7, 55
statistics, viii, 26, 50, 55, 56
storage, 12, 45
strength, 24
summer, 39
supply, vii, 1, 2, 3, 4, 7, 12, 13, 16, 18, 19, 25, 31, 37, 38, 44, 48, 52
supply chain, 3
supply disruption, 2, 4, 37, 52

T

targets, 11
tax rates, 42
taxes, 3, 39, 41, 46
technological developments, 18
temperature, 2, 3, 21
tension, 12
testimony, vii, 5, 7
Texas, 44
threatened, 38
threats, 3

Three Mile Island, viii, 26, 43
time frame, 11
timetable, 43
timing, 13
total energy, viii, 25, 26, 28, 30
trade, 4, 55, 56
trading, 40
transformation, 2
transparency, 17
transport, 36
transportation, vii, viii, 25, 26, 27, 28, 31, 32, 36, 47, 48, 51
trucks, 3, 36, 52

U

U.S. Geological Survey, 15, 16, 22, 59
uncertainty, 13, 23
United Kingdom, 22
United States, vii, 1, 2, 3, 4, 15, 17, 19, 21, 22, 25, 31, 33, 38, 42, 48, 59
upload, 42
uranium, 24
Utah, 20

V

vehicles, 51, 52
vulnerability, 2

W

water, 10, 12
wells, 9
Western Hemisphere, 31, 33
wholesale, 56
wind, viii, 7, 22, 26, 27, 51
wood, 7
Wyoming, 20, 21